AF866702

Atlas of Chinese Macrofungal Resources

Yu Li • Taihui Li • Zhuliang Yang • Tolgor Bau
Yucheng Dai

Atlas of Chinese Macrofungal Resources

Volume 3: Cantharelloid Fungi & Agarics

Springer

Central China Publishing Group

Central China Farmer's Publishing House

Yu Li
Engineering Research Center of Chinese Ministry of Education for Edible and Medicinal Fungi
Jilin Agricultural University
Changchun, China

Zhuliang Yang
CAS Key Laboratory for Plant Diversity and Biogeography of East Asia
Kunming Institute of Botany, Chinese Academy of Sciences
Kunming, China

Yucheng Dai
Institute of Microbiology
Beijing Forestry University
Beijing, China

Taihui Li
Guangdong Provincial Key Laboratory of Microbial Culture Collection and Application
Guangdong Institute of Microbiology, Guangdong Academy of Science
Guangzhou, China

Tolgor Bau
Key Laboratory of Edible Fungal Resources and Utilization (North), Ministry of Agriculture and Rural Affairs
Jilin Agricultural University
Changchun, China

ISBN 978-981-96-0501-9 ISBN 978-981-96-0502-6 (eBook)
https://doi.org/10.1007/978-981-96-0502-6

Jointly published with Central China Farmer's Publishing House

The print edition is not for sale in China (Mainland). Customers from China (Mainland) please order the print book from: Central China Farmer's Publishing House.

© Central China Farmer's Publishing House 2025

This work is subject to copyright. All rights are solely and exclusively licensed by the Publisher, whether the whole or part of the material is concerned, specifically the rights of reprinting, reuse of illustrations, recitation, broadcasting, reproduction on microfilms or in any other physical way, and transmission or information storage and retrieval, electronic adaptation, computer software, or by similar or dissimilar methodology now known or hereafter developed.
The use of general descriptive names, registered names, trademarks, service marks, etc. in this publication does not imply, even in the absence of a specific statement, that such names are exempt from the relevant protective laws and regulations and therefore free for general use.
The publishers, the authors, and the editors are safe to assume that the advice and information in this book are believed to be true and accurate at the date of publication. Neither the publishers nor the authors or the editors give a warranty, express or implied, with respect to the material contained herein or for any errors or omissions that may have been made. The publishers remain neutral with regard to jurisdictional claims in published maps and institutional affiliations.

This Springer imprint is published by the registered company Springer Nature Singapore Pte Ltd.
The registered company address is: 152 Beach Road, #21-01/04 Gateway East, Singapore 189721, Singapore

If disposing of this product, please recycle the paper.

Foreword 1

Species are carriers of genes. A gene existing independently from individuals within a species has no value. A diversity of species incorporating a diversity of genes existing within a diversified ecosystem represents the core of biological diversity. Genetic diversity of genes does not exist without diversity of species. Therefore, species diversity is the most important reproductive natural resource that man has and is a treasure that humankind relies on to achieve sustainable development.

Fungi are one of the most diverse biological groups in the earth biosphere, and in term that refer to all eukaryotic fungal life forms, including the true fungi from Kingdom Fungi, chromistan fungal analogues such as the oomycetes, as well as protozoan fungal analogues such as myxomycetes. Little has been known about the diversity of fungi. A conservative estimate by specialists suggests that there are at least 2.5 million species of fungi among fungal species within the biosphere on the earth. However, only around 100,000 species have been known and denominated, leaving 96% of fungi to be discovered, recognized, denominated, described, studied and utilized. Like other microbes, fungi can be produced in factories at a large scale and developed into a large industry through the application of technology, perhaps making it the most abundant reproducible natural resource that humans count on for their sustainable development.

Chinese Fungal Resources and Exploitation, a collection of four volumes, that includes *Atlas of Chinese Macrofungal Resources*, *Production of Edible Mushrooms in China*, *Mycomedicines in China* and *Processing of Edible Mushrooms in China* is the most comprehensive compilation and collection of fungal resources and their exploitation in China today. Renowned mycologist, Academician Yu Li served as the editor-in-chief.

More than 1800 species of macrofungi are presented in the volume *Atlas of Chinese Macrofungal Resources,* which are the original works of the authors of this book; many are newly published, representing the latest achievements in the macrofungal researches. This volume is characterized by the concise descriptions, exquisite images and strong practicability in format that makes it an important referential tool for identification of fungal resources.

The volume *Production of Edible Mushrooms in China* systematically introduces various cultivation technologies used in the production of edible mushrooms, the fifth largest agricultural crop in China, and includes successful cases and experiences in the cultivation process.

The introductory part of the volume *Mycomedicines in China* introduces the definition, origin, development and herbalogical textual research of mycomedicines; their medicinal properties and compatible theory in traditional use, chemical composition, pharmacological action, identification, production and ethnic mycomedicines are also presented. And the following individual chapters introduce the mycomedicines, respectively, under various categories of stroma mycomedicines, sclerotium mycomedicines, fermentation mycomedicines and others.

The volume *Processing of Edible Mushrooms in China* introduces the current situation and prospects of processing fungi and covers aspects of preservation, logistics, facilities, equipment, primary processing, complex and testing of processing quality. Examples are also given.

Sun Simiao, the ancient Chinese King of Medicine, classified the human health status into three levels: superior, intermediate and inferior, in which, free-of-illness (healthy) is superior, close-to-illness (sub-healthy) is intermediate and illness-possessed (unhealthy) is inferior. In the same way, he classified medical treatments into three levels: making humans free from illness is superior, curing almost-ill people is intermediate and curing ill people is inferior. His healing art demonstrated the broadness and profoundness of Chinese medicine and herb use long before the establishment of the theoretical system in which the prevention of illness prevails over the treatment of illness.

Modern science has proven and will continue to prove that edible mushrooms have great significance in maintaining a superior status of health in man. Therefore, while great efforts are being made into developing edible fungus industry, it is undoubtedly critical to carry on the research, development and exploitation of edible mushrooms in a broader and profounder way, while collaborating with medical and health systems, and incorporating industry, education and research sectors, in order to capitalize the achievements of the present era and create benefits for generations to come. The publication of the four-volume collection of *Chinese Fungal Resources and Exploitation* will provide guidance and reference for research, development and exploitation of edible mushrooms by incorporating the industry, education and research sectors, as well as creating the foundation for the development and innovation of fungi for the benefit of mankind.

Jiangchun Wei

Academician of Chinese Academy of Sciences
Honorary chairman of the Chinese Society of Fungi
Editor of the Editorial Committee of Chinese Spore
Flora, Chinese Academy of Sciences
Researcher of State Key Laboratory of Mycology
Institute of Microbiology, Chinese Academy of Sciences
Beijing, China

Foreword 2

When I received the invitation for preparing this foreword, I considered it as a great challenge, as for years, I have rarely written letters or articles in Chinese. While I was still hesitating, something said by Dr. Sam Nujoma, former president of Namibia accidentally came to my mind: "I always take challenges as opportunities to learn new things". Indeed, I did run into many difficulties and challenges when writing this foreword, especially typing in Chinese using Pinyin (a romanization system for Standard Mandarin Chinese). On the other hand, I did learn a lot of new things because of the opportunities offered by this challenge.

Mushroom (edible fungus) biology, a new branch of mycology, is specialized in the discussion of mushroom morphology, function, genetics, evolution, ontogeny, exploitation and its basic relations with the environment. Mushroom biology differs from mushroom science. Mushroom science is a branch of mushroom biology, as it mainly involves the mushroom cultivation, its production principles and practice. Mushroom biotechnology is a part of mushroom science and mainly involves mushroom products produced by fermentation or extraction. Although these discussion topics do not change, the methods of studying change with each passing day as natural science develops. As a result, the contents in textbooks on mushroom biology, research topics and methods are frequently added, deleted or updated. The collection of *Chinese Fungal Resources and Exploitation* having been compiled on the foundation of science, technology and research represents the most recent compendium of mushroom biology (macrofungi) in China. Many have underestimated the policies and results of science, technology and creation (STC) of edible fungi in China. Statistics shows that the power of investment in mushroom research, development and exploitation is substantial, which will inject energy into the continuous expansion of basic research and industrial development of edible fungi.

A clear-cut explanation on the definition of edible fungus (mushroom-macrofungus) has been given by the author in his preface. The term "edible mushrooms" or "medicinal mushrooms" is used in most international conferences and articles related to macrofungi, while edible fungi or medicinal fungi is rarely used.

It has been estimated recently that there are approximately three million fungal species on earth (Hawksworth DL, 2012. Biodivers Conserv 21:2425–2433; Wasser S P, 2014. Biomed. J. Sci. 37:345356), among which, about 100 thousand true fungal species have been denominated in 2012. However, new species continued to be discovered. In the past 10 years, around 60% of the new species of fungi were found in tropical climates. Up to now, it is estimated that there are 150~160 thousand mushrooms on earth, with approximately 16,000—only 10% of that are known. Among these, about 2000 are considered safe and edible, including over 700 mushrooms that have medicinal values. The biodiversity of mushrooms is a comprehensive biological science, which is of great importance for the investigation, identification and exploitation of new mushroom species in the future. The many issues that surround biodiversity are highly complicated. For example, how to build up protection policies including an understanding of the interaction between genetics and taxonomy, and how to explore new edible and medicinal mushroom resources from different levels. Therefore, it is absolutely critical to protect and genetically conserve pure wild species germ plasm.

Despite the fact that modern technology is playing a more and more important role in human civilization, the well-being of today's man is still confronted with three great challenges: regional food shortages, the decline in health quality as well as the deterioration of the ecological environment, which will become more and more serious as the population continues to grow worldwide. We hunger for the use of fair and effective global knowledge and technology to solve or alleviate these three challenges, especially the occurrence of harm to human health, which is not limited to poverty-stricken countries or a poverty-stricken class of a society. As a matter of fact, people living in well-educated and wealthy families in both developing and developed countries have many health problems as well, such as high blood pressure, heart disease and cancer—the so-called affluenza. These health problems bring unhealthy economic results, increase the living cost for both consumers and taxpayers, weaken labour capacity, which then becomes the main reason for low productivity.

Neither an individual nor a country can ignore this issue. I believe deeply that mushrooms can make great contribution to the three great challenges human beings face. Having been engaged in mushroom education and research for over 50 years, I have had opportunities to spread knowledge about mushroom biology, its research standards and utilization across five continents. Mushrooms are not only able to convert biological wastes containing copious cellulose and lignin into food but also produce medically and nutritionally valuable products of great significance to human health. The most remarkable advantage of mushroom cultivation is that once managed properly, it will alleviate the deterioration of the ecological environment, and probably even bring pollution of the environment to zero. Meanwhile, the formation of a foundation and subsequent development of mushroom industry offers new employment opportunities. Furthermore, cultivating and developing edible and medicinal mushrooms can create positive economic growth and have positive influence on individuals, regions and economic growth of countries. Therefore, mushroom research and development will continue to expand in the future. Since mushrooms produce (the mushroom itself), mushroom products (mushroom derivatives) and mushroom protection (utilization of waste) will contribute to the three challenges human beings face, sustainable research and development of mushroom research and exploitation can become a "none-green organisms" (as edible fungi are none-green organisms that do not contain chlorophyll).

In general, Professor Yu Li, the editor-in-chief, has taken great efforts in this collection from its conception to implementation. Authors of this collection are all extremely experienced in mushroom education and research. This is a set of magnum opus that comprehensively, completely and systematically introduces theory on mushroom resources, taxonomy, production and processing in China, a valuable masterwork on mushroom science.

S. T. Chang

Emeritus Professor, Department of Biology
The Chinese University of Hong Kong
Hong Kong, China
Written in Canberra, Australia
January 10, 2015

Preface

After 3 years of effort, *Atlas of Chinese Macrofungal Resources,* the first volume of *Chinese Fungal Resources and Exploitation* has been completed. Upon sending to press, here are a few words from my heart to the readers.

Fungal groups covered in this book are the large species mainly composed of mushrooms. Mushroom is also called gujun (Chinese) or macrofungus, so as to be distinguished from other microfungi such as mould and yeast. Chinese ancient texts named macrofungi grown on wood as Xun, and those in soil Jun, as commonly called mogu. In fact, originally the Chinese character "菌" (pronounced as jun) specifically refers to this type of fungi, an equivalent to the English word "mushroom", German "Pilze", French "champignon", Russian "грибы" and Japanese "キノコ". As documented in *Origin of Chinese Characters* (*Shuo Wen Jie Zi,* first Chinese dictionary, compiled by Xu Shen, 121 AD), the Character "菌" indicates that it is a life being that is of a plant nature, with a shape of a round granary. There are over 30 life beings of this type recorded in ancient Chinese documentations. The common name "蘑菇" (mushroom) only started to be used since the Yuan Dynasty. The definition of such life being by *DICTIONARY OF FUNGI* is: "a macrofungus with a distinctive fruiting body which can be either hypogeous or epigeous, large enough to be seen with the naked eye and to be picked by hand", which refer to *Cordyceps* spp., *Morchella* spp., *Peziza* spp., *Tuber* spp. (truffle) of Ascomycetes; agarics, puffballs, boletes, coral fungi, Lingzhi and jelly fungi of Basidiomycetes. On the other hand, fungi like larger pathogenic fungi and myxomycetes can both "be seen with the naked eye and to be picked" (large enough), and there is no lack of atlas of macrofungi in the world that includes them.

To make this book a good one, authors had to consider these questions:

First, Why This Kind of Book?

There are a lot of records on fungi in vast ancient Chinese documentations, and a systematic atlas started in the Northern and Southern Dynasties (420–589), with no less than a thousand publications until today. Especially in the past 10 years, new books started mushrooming, brought forward by the press of different regions and types. The value of spreading science by way of these books is tremendous, not to mention the great efforts by the authors. They have made fungal biology previously overlooked more and more interesting. Nevertheless, among existing publications of the like because of the scientific knowledge of the time when published and the authors' level of expertise, inaccurate documentation can often be seen. As the science and technology advance and the Chinese national power strengthens, the people of the country have become leaders in this field of study. While great number of new species are discovered, many identification errors in the past have been corrected, and recognition of the vast fungal resources in China has been refreshed. Therefore, authors and many peers realized the necessity to publish an atlas of macrofungi that represents the contemporary research level in China, that is on a par with internationally excellent work and that can comprehensively and scientifically portray resources in China based on the accumulation of research results in the past 10 years (especially nearer years) and provide tools to correctly identify the macrofungal resources in China. Without the right tools, people might be doing work in a less efficient way, somewhat like watering the garden by carrying a huge jar as Chinese said in the past, resulting in mistakes that might pass on even to our children. Therefore, the publication of this book is the call of the era and the need of the industry, as well as the expectation of our peers.

Second, What Kind of Book?

Books on mushrooms around the world, from monographs on different taxa to popular science readings, are numerous and even too many to catalogue. Monotonous are the books of the like in China in content, structure and layout, with some having identical contents, excerptions and even copies from others, but few are stunning. When writing this atlas, the authors had taken into account the following hypothetical scenario: should it become a voluminous work on taxonomy, then only a few readers would be able to understand or appreciate it; on the other hand, should it become entry-level book, it is easy to get stuck in a rut and lack new ideas. Therefore, authors strived for a breakthrough on contents, materials, layout and editing to make both pictures and text excellent, scientifically accurate, easy to understand, elegant and vivid. "Admired by scholars and laymen alike, welcomed by most readers, used in wide" and "popular but not vulgar" were the common views and the target to pursue by authors.

Third, What Style?

How can this book serve both as references to a scientist and entry level readings to young people, while enjoyed by many? Neither obscure scientific terminologies nor systematic arrangement style of a taxonomist would work. Therefore, when introducing fungal species, images and texts were not arranged strictly following the arrangement scheme popularly adopted by the modern taxonomy system, but starting with morphology as the main recognition characteristics, and then divided into ten big groups. Species in the same group were arranged by alphabetic order of their Latin names for easy reference; generic name of one variety that matches recent taxonomic research results or molecular biology evidence is used when assigning it to a generic group. The systematic relations of all macrofungi of each genus are sufficiently illustrated in the part Overview. This way, it is not only convenient and straightforward but also reflects the connotations of modern scientific results.

Relevant knowledge is manifested in overview and glossaries. All is made easy to start, grasp, understand and use, which brings a vividness to this academic book of knowledge, without becoming serious and monotonous.

Fourth, How to Demonstrate the Scientific Value of This Book?

Scientific value is undoubtedly the soul of all scientific writings, but science is a progressive knowledge system. As the level of knowledge advances, scientific methods improve, and knowledge of the interrelation among things gets deeper, therefore the continual correction of mistakes becomes inexorable. On a certain space-time coordinate point, relative accuracy is the embodiment of scientific value. This is the realm the authors pursue. Therefore, to be factual and scientific, geographic zones of macrofungal resources defined in this book have been amended according to geographic and natural ecologic environmental factors based on the research of our predecessors, rather than some administrative division. What is used are the most recent research results based on the vegetation classification in China in the regionalization system in this book, combined with specimen collection records in the past, and regional distribution consideration is given to relevant species. Hopefully this new attempt will present the fungal resource characteristics in all locations within China and their relationship with natural climate, geography, habitat and ecology in a more reasonable, scientific and factual manner. Similarly, descriptions and images in this book are based on the confirmed results of specimens studied, compared with those of the same types or groups both home and abroad and then identified by authors, strictly following taxonomical principles. None have been selected by resemblance. Assumption based on uncertain information or identification by pictures was ruled out. During compilation of this book, the following principles were adhered to: candidates with uncertainty out; pictures, specimens and description that are not consistent or do not corroborate each other out; images unclear, unreal or imprecise out. Even if affiliation of any changes in the future, the specimens and evidence still remain. Authoritativeness is manifested through scientific value, with only truth is talked about, which reflect the saying "practice is the sole criterion of truth".

Fifth, How to Reflect the Representative Value of This Book?

As an atlas covering the main ecological regions across the country, its extent and representative value are consistent. Generally speaking, only things coming from China can represent China; the more taxa come from a country, the better they demonstrate the characteristics of that country; only by including more taxa can each biological taxon be better represented; only experts who have studied and recognized many species of certain taxon and gone deeper into their biology can better represent the knowledge level of that taxon. To sufficiently bring forward the representation, more than 1,800 species (variants) of fungi were selected from millions of specimens collected from different ecological regions of China to be included in this book, making it an atlas of macrofungi that includes the most numerous species, denominated by Chinese scholars, and most numerous species discovered with authors' personal involvement of the like in China until today. Since China has abundant and diverse fungal resources, with many extremely similar to those abroad, unexpected difficulties always exist when identifying a specimen accurately. A great number of specimens are either questionable or lack of reliable proof. Some specimens are unavailable though they have been recorded. Such paucity, regretfully, will have to be passed on to future generations. Fortunately, the four main co-authors have been working closely with me and made great contributions to the scientific value and representative value of the atlas. All of them are frontline scientists, young and full of energy, who have intact education or research experience in both home and/or abroad, and who have done research in depth in each of their own fields. During the compilation of this book, the portions they each undertook were fields particularly familiar to them, and who represent the current cognitive level of relevant macrofungal resources in China. Great support and assistance were offered by experts from different subfields throughout the country, resulting in a greater representative value of this atlas.

Sixth, How to Present the Originality of This Book?

The lure of originality is irresistible to scientists who are not willing to be outdone, and so it is expected by the readers. The originality of a book is vital. Therefore, it was the objective of the authors devoted in their own field for many years to maintain originality and respect for the reader, by including pictures and specimens of fungal taxa which had been verified repeatedly. This is the presentation of their academic achievements accumulated from long-term research efforts. Quality was put before quantity while selecting materials, with solid proof accompanied and requiring that each must be clear, accurate and honest. Confusion, unconformity or poor quality were not tolerated. All of the images and pictures came from China, and all descriptions were made based on Chinese specimens (including hundreds of type specimens that the authors of this book used when publishing new species) and pictures, of which quite a number are endemic species and newly recorded species through many years of research by authors. None was borrowed from overseas, none thrown in to make up the number or plagiarized! The result is to guarantee the originality and truthfulness of their work.

Seventh, How to Present the Contemporary Value of This Book?

Species, relatively stable, independent to a changing life system are the most basic units for the understanding of the world of life through science, and which is formed through reproductive and geographic isolation throughout the long-term evolutionary process. Innumerable biologists are fascinated by the mystery of biological diversity. The accumulation of knowledge by generations of taxonomists has created a valuable resource. On top of it, the development and application of modern molecular biology has continued to improve people's knowledge of the species concept. The newest discoveries of science have made the evergreen tree of life sparkle brightly. For historical reasons and for those that we all know, in the past 200 years, China's domestic scientific research on her resources has fallen behind that of Japan and the Western world. Even today, quite a few areas of research lag and must refer to, imitate or even borrow data from beyond China. China's knowledge of its own fungal resource is no exception. Due to all limitations from the past, a lack of knowledge about species as denominated by the West made it difficult to compare them with specimens of our own, therefore resulting in the inability to conduct research on our own specimens in depth, hence a creating significant existing confusion of names. Many of China's species have been mistakenly identified as species reported by developed countries in Europe and America for a long period of time, causing tremendous confusion to science, production and daily life. In fact, the differences are great between biological taxa in the east and the west by nature. Here is one case. Lingzhi, highly valued in China for thousands of years, and still popular and beloved today, has long been titled *Ganoderma lucidum*, a name from Europe. Indeed, the commonly cultivated Lingzhi in China is anything but *Ganoderma lucidum* commonly seen in Europe; neither is the widely cultivated black wood-ear in China the same species as the European *Auricularia auricula-judae!* Let alone Xiuqiujun, the "rising star" of today, is absolutely not the *Sparassis crispa* in Korea and Japan but *S. latifolia*. Many are the cases like these. Since the reforming and opening of China, the authors have discovered a large number of new fungal species and new resources in each of their respective field of specialty and have corrected many mistakes made in identification. This atlas, to a great extent, has summarized the research results of the recent decades in China, bringing forth the most recent research results in a contemporary way. For example, species introduced in this book that have been published after the year 2000 alone are as many as 380—all showing an impress of the new epoch.

Finally, How to Present the Artistry of This Book?

A good book that becomes a must-have for scientific and technological workers, a book that readers hate to put down, in addition to its novelty, richness and splendidness, the artistry is of great importance. The quality of editing, layout, paper, binding and printing are important factors, none of which can be poor or absent. The saying "fine feathers make fine birds" explains it all. Of course, "fine feather" should not eventually become the apple of Sodom. Editing and publishing, a profession of its own, can completely affect the level of a publication's quality both inside and outside. Publications of similar type, in comparison, can vary in quality. In terms of being exquisite, publications in Japan and Germany enjoy a good reputation. How to make this book equivalent to those is a severe test to the publisher. With colour printing equipment standardized in quality today, factors such as layout, paper and printing represent the vision, level and handling of a publisher to a greater degree. I am very grateful to the director, editor-in-chief and their colleagues of Central China Farmer's Publishing House for the great effort they put into this project. Despite their history of producing multiple splendid presentations, once again they strived for perfection, and even went off the beaten track, to achieve more visual impact and aesthetic effect in the layout: a combination of science and art, a presentation of culture and beauty! I believe compliments naturally burst forth when readers are presented with this new book, while a delicate smell of printing oil is still given off. We also believe that time will confirm the scientific and artistic value of this book.

Besides the hard work by the authors of this book, generous contributions were offered by some renowned research organizations and scholars from both home and abroad, their supply of data, specimens and pictures—icing on the cake. My gratitude goes to them all!

Nowadays, many in academia no longer bother to recognize life forms at morphological level! Not to mention field trips! Hot is the game of four base-pairs! Nothing is wrong with doing research using new technology itself since through history it is inevitable that traditional science continues to fuse itself with new technology. But favouring one over another, or even abandoning the fundamentals, is just like the blind men trying to size up the elephant-taking a part for the whole and building a tower upon sand. Such research behaviour is a misguidance to future generations! By surrendering to SCI impact factors, many go after fame and gain, like flies after a stink, vying for the higher, while becoming flighty and impetuous. What a vanity fair! Can we call it Sodom and Gomorrah of academia? Who is able to save it then? Fortunately, some unreconciled intellects stay sober, sane and simple, with an attitude of "seek root, leave flower to others". They care only for concrete matters, spending most of their time in the field, where the main battlefield of economy remains—a ray of hope.

"I oft regret spring's gone without leaving its trace; I do not know it's come up to adorn this place". This Chinese ancient poem keeps reminding us that hope is always around. After dozens of years of travelling over mountains and rivers, exposed to cold and wet, looking through every tree and bush for a trace of mushroom, researchers have been able to present these lovely beings, with freshest looking view: many littles make a mickle. Publication can mean an Omega, but it can also mean another Alpha to newer, higher and finer chapter yet to come! Young mycologists, play not the old music, but compose your own!

Neither do we have the intention nor the nerve to claim that this book is perfectly accurate, but we did our utmost to bring errors to a minimum level. We expect it to be the new starting point for the mycological resource research in China. A long journey can be covered only by taking one step at a time, and an ocean can be made only by saving every drop of water. Mycologists, bear this in mind and start from within yourself now, or you will not be able to make steady progress and innovation like the waves behind driving on those before.

Move on and we look forward to your great achievements!

With compliments,

Changchun, China
Winter of 2014

Yu Li

User's Guide

Structure

1. Overall structure

   The overall structure is composed of forewords, preface, table of contents, overview, colour photographs and descriptions of all species of the macrofungi, glossaries, references, indices of Chinese and Latin scientific names, afterword and acknowledgement.
2. Overall division

   Overview, colour photographs and characterization of every species of the macrofungi make up the main body of this book. The 1800 plus macrofungi are divided into 10 groups and then introduced each according to their taxonomic and morphological characteristics.
3. Introduction of species

   The introduction to each species typically includes at least one colour photograph in its natural environment, its Latin scientific name, its main macroscopic and microscopic structural characteristics, ecological habits, distribution and economic value, etc.

How to Search

Several methods have been provided for easy access to a specific species or group depending on the choice of the reader.

1. Search by Latin scientific name index (in alphabetic order)

   The page number of text where certain species appears can be searched by the Latin scientific name index at the back of this book if known, through which further information on this species can be accessed. The same method applies if Latin synonyms are available.
2. Search by Chinese name index (in Chinese pinyin's alphabetic order)

   The page number of text where certain species appears can be found by the Chinese name index at the back of this book if known, through which further information on this species can be accessed. The same method applies if Chinese vernacular names are available.
3. Search by table of contents

   All species belonging to different groups are arranged in alphabetic order by their Latin scientific names. If the Latin scientific name of a certain species and to which of the ten groups it belongs is known, the page number of text where this species appears can be accessed by looking up table of contents of this group. For example, when searching for the page number of text where *Agaricus bisporus* under the group of Agrics appears, by looking up to the part of "AGARICS" in table of contents, one can find that the page number of text where *Agaricus bisporus* appears.

4. Search by colour blocks in specific parts on side cut and bottom cut

   Different colour blocks are printed above even page numbers in the lower left corner and to the left of the odd numbers in the lower right corner in this book to make distinguishing different parts easy is for overview, is for the larger ascomycetes, is for the jelly fungi, is for the coral fungi, is for the polyporoid, hydnaceous, and thelephoroid fungi, is for the cantharelloid fungi, is for the agarics, is for the boletes, is for the gasteroid fungi, is for the large pathogenic fungi on crops, is for the large myxomycetes, is the glossaries, is the indices and is for the reference.

On Partial Formats and Contents

1. Names of fungi

   Normative Latin names of each fungus are given in the detailed introduction on different species of macrofungi in all Chapters. Frequently used Chinese vernacular names will be listed in brackets after each fungus' normative Chinese name if available; frequently used and important Latin synonyms will be listed under the fungus' scientific name if available.
2. Norms of scientific names

   Norms used for scientific names in this book is mainly based *on International code of Nomenclature for algae, fungi, and plants,* cross-checked with *Index Fungorum, Dictionary* of *Fungi 10th Edition,* literature where the scientific name first appeared, textual research papers on scientific names and the most recently published literature on mycological taxonomy. This includes the principle that the norms for authors (nomenclators) must comply with the regulations, for example, abbreviation of nomenclator contained in a scientific name must comply with regulations of *Authors of Fungal Names* and *Authors of Plant Names.*
3. Omission of nomenclator

   When the number of authors (nomenclators) of a scientific name is 1~2, all names will be listed, for example, both nomenclators "Zhu L. Yang & T. H. Li" in *Amenita exitialis* Zhu L. Yang and T. H. Li are listed; when the number is 3 or above, only the name of the first nomenclator is listed and the ones behind will be replaced with "et al.", for example, there are three nomenclators, "W.Q. Deng, T.H. Li & Zhu L. Yang", for *Amanita macrocarpa* W.Q. Deng et al. indeed, but it is shown in this book as "W. Q. Deng et al."
4. Simplification and abbreviation of scientific names

   To save space, except for the purpose of giving an introduction on a specific fungal species itself, only names of the genus and specific epithet is listed while all nomenclators are omitted for the scientific name of a fungus listed for discussion or example illustration; the generic name is given in full when it first appears in one part, but will be abbreviated (i.e., the first letter of the generic name is capitalized with a foot a dot) when it appears again, then followed by specific epithet and so on.
5. Size of fungus in image

   For the purpose of typography and reading, the size of images of fungi is neither reduced nor enlarged to a consistent scale according to their actual size. Description on a fungus should be referred to for its actual size only.
6. Order of images

   For easy access to main fungi in this book, image order in Chapter 1 of Volume I Overview is arranged separately; serial number of each species is consistent with that of its description.
7. Order of references

   References in Chinese and foreign languages are listed separately. References in Chinese are arranged in pinyin alphabetical order and those in foreign language in English alphabetic order.

8. Morphologic characteristics

Morphologic characteristics are presented by both colour images and text descriptions. Macroscopic characteristics of a fungus by text description include size, shape, colour (and discolouration because of a bruise) and appendant of its main parts. Microscopic characteristics include size, shape and surface features. Major changes to colour and size of a fungus might occur because of the impacts of its ecologic environment, weather and maturity, which may not be entirely covered by images or text description. Some characteristics in common with or similar to fungi of the same type are usually omitted in text descriptions.

9. Distribution

Distribution area of macrofungi in China is divided into seven major regions in this book: Northeast, Northern, Central China, Southern China, Inner Mongolia, Northwest and Qinghai-Tibet Region. Introduction on each species contains its distribution information in other regions. Some reports on the distribution of certain species were made on the basis of mistakenly identified specimens or even without availability of a voucher specimen. Although a small number of species have been reported in certain regions of China because the authors of this book were not able to find solid and reliable confirmation, distribution of these species has not been included in this book. Therefore, known distribution range listed in this book may seem narrower than that introduced in other literature.

10. Economic values

Brief introduction on known edible, medicinal and economic values of different species is provided in this book; no notation is given to the species with unclear economic values.

11. Conventional expressions

Conventional expression accepted and commonly used by international mycological colleagues for size and measurements of macrofungi's microscopic structure is adopted by this book, i.e. the number before "×" indicates the length of the structure and width behind, with length unit appearing only once at the end. For example, when indicating the length of a spore as 9~10 μm and width 4~5 μm, the form "spore 9~10 × 4~5 μm" is used for description, rather than the form "(9~10) g × (4~5) g" as required by national standards. This is to be in unity with the international mycological circle and easier for exchanges among colleagues. In addition, some conventional expressions in this book, for example, "pip-shaped" means the shape of apple or pear seeds, rather than shapes of other seeds; "pinkish" means the skin colour of a regular white person or a yellow person with pinkish colour. These notes are for the correct understanding.

Mushrooms and toadstools and other larger fungi have become increasingly popular in recent years. They are not only of fascinating scientific interest but are extremely beautiful and variable in their form, offering pleasure to both artist and photographer. Their "overnight" appearance has made them objects of mystery and folklore for centuries. In addition, many species are edible and delicious, so providing a ready source of food and enriching many culinary dishes.

D. N. Pegler

Acknowledgements

Strong support from many colleagues at home and abroad was received for the writing of this book. Special thanks to:

Academician Jiangchun Wei of the Institute of Microbiology, Chinese Academy of Sciences, and Prof. S. T. Chang, Lifetime Honorary Professor of the Chinese University of Hong Kong for the forewords of this book.

The leader, editors, designers and distributors of Central China Farmer's Publishing House have made great efforts for the publication of this book. The Henan Provincial Institute of Remote Sensing and Mapping assisted in the production of the geographical zoning map of macro fungal resources in this book.

The Fungarium, Institute of Microbiology, Chinese Academy of Sciences (HMAS), Cryptogamic Herbarium, Kunming Institute of Botany, Chinese Academy of Sciences (HKAS), Herbarium of Microbiology Institute of Guangdong (GDGM), Herbarium Mycology of Jilin Agricultural University (HMJAU), Herbarium of Institute of Applied Ecology, Chinese Academy of Sciences (IFP) in Shenyang and Herbarium of Beijing Forestry University (BJFC) provided a large number of detailed and supporting specimens for this book.

The Royal Botanic Gardens, Kew (K), the Herbarium, Institute of Botany, University of Vienna, Austria (WU), Farlow Reference Library and Herbarium of Cryptogamic Botany, Harvard University (FH), and the Herbarium, Department of Botany, National Science Museum, Tsukuba, Ibaraki in Japan (TIVS) provided the authors great support for studying relevant model specimens and authoritative specimens.

When writing the content about fungal geography and ecological environment, we gained the help from Researchers Dianxiang Zhang, Fuwu Xing and Hongfeng Chen of the South China Botanical Garden.

Sincere thanks to all the friends who helped in the preparation and publication of this book!

Brief Introduction

One thousand eight hundred and nineteen species (or varieties) in 509 genera of macrofungi known from China are documented in this work. According to their morphological characteristics, they are practically divided into 10 groups and then introduced accordingly, including 196 larger ascomycetes, 21 jelly fungi, 47 coral fungi, 637 polyporoid, hydnaceous and thelephoroid fungi, 11 cantharelloid fungi, 653 agarics, 130 boletes, 75 gasteroid fungi, 16 larger pathogenic fungi on crops and 33 larger myxomycetes. All species are evidenced with vouchers and photographs. About 370 species (occupying one-fifth of the total species) with type localities in China are included, among which over 260 species (accounting for one-seventh of the species) were firstly discovered and published by the present authors, some of them as specific species in China and East Asia, which have tried to present the latest knowledge about the Chinese macrofungal resources.

All species are described accompanied with colour photographs showing their macromorphology and (or) habitat. The macroscopic and microscopic diagnostic characters, ecological habits, economic importance (edibility, medicinal availability or toxicity) and geographical range in China are provided. The characteristics and using method of the book, related mycological vocabulary, common taxonomic techniques and positions of the fungal genera in modern taxonomic system are briefly introduced. For the convenience of the readers, all species in the different parts of main text are arranged in alphabetical order of their Latin names, and indices of Chinese names and scientific Latin names to all species are appended.

The knowledge of this book should be interesting for mycologists, mycology fans and mushroom lovers, as well as for researchers, teachers and students studying on edible fungi, plant pathology, healthcare and biomedicine sciences, bioresources and biodiversity, ecology and other related disciplines. It is an ideal reference for those who are interested in the Chinese macrofungi and larger slime moulds.

Foreword to the English Version of the Atlas

In the last century, the understanding and research level of macrofungal resources in China lagged far behind that of developed countries in Europe and North America. Not only was the number of known species fairly limited, but misidentification was rather common, with quite a few species endemic to China or Asia being misidentified as those already described in Europe, America, or elsewhere. There were only a few published atlases of Chinese macrofungi at that time.

Since the beginning of this century, China has rapidly developed in the field of mycology and has gradually become the most active country globally in mycological research, with the highest number of new species discovered each year. The *Atlas of Chinese Macrofungal Resources* was compiled by several of China's most active research teams on macrofungal resources, showcasing numerous discoveries in this century. The Chinese version of the atlas was published by Central China Farmers Publishing House in 2015.

Compared with the previous macrofungal atlases in China, the *Atlas of Chinese Macrofungal Resources* contains the most species, including 1819 species of 509 genera. Among them, more than 330 new species were discovered in this century, reflecting the latest research results. Over 370 species originally described from China were included, better reflecting the characteristics of China's macrofungal resources, with more than 260 species jointly discovered by the five lead authors and their collaborators. Since its publication, the atlas has been warmly welcomed by the majority of readers, affirmed by domestic mycologists and other readers, and has also attracted high attention from overseas enthusiasts.

However, being an atlas compiled in Chinese poses a challenge for most foreign readers who do not understand the language. Therefore, Springer Publishing House, Central China Farmers Publishing House, and the main authors of the Atlas reached a consensus to translate the Chinese content into English for publication, enabling more foreign readers to comprehend the content.

More than 8 years have passed since the atlas was first published. During this period, research on macrofungal resources in China has accelerated, with a continuous discovery of more new species. Upon comparison with recent research results, it was found that some pictures in the atlas represent new species published in recent years after its initial release.

As a translated work, maintaining consistency with the original content is crucial. For species requiring name changes in light of recent research progress, the original names are retained. However, the main authors of the atlas and the translators also seek to use the opportunity of publishing the English version of the atlas to inform readers of the name changes for these fungi, providing more insight into recent research progress in China. The names of the species that should be changed are listed in the following tables, with the hope that it will aid readers in understanding these fungal species more accurately.

The task of determining whether the species names in the atlas should be changed is highly professional and time-consuming. To accomplish this, in addition to several main authors of the original work, many younger mycologists participated. They are (in alphabetical order by name): Qing CAI, Baokai CUI, Yuguang FAN, Guojie LI, Jipeng LI, Ting LI, Junfeng LIANG, Chaoqun WANG, Xianghua WANG, Yuanbing WANG, Fang WU, Gang WU, Niankai ZENG,

Ming ZHANG, Ping ZHANG, Ruilin ZHAO, Changlin ZHAO, et al. The authors and translators express their heartfelt thanks to them.

The leaders and editors of Springer Publishing House and Central China Farmers Publishing House have provided the necessary publishing funds and contributed significantly to the publication of the English version of the atlas. The authors and translators express their sincere gratitude for their enthusiastic support and assistance in various aspects.

Due to the limited knowledge and English proficiency of the translators, and considering the extensive nature of the translation content, errors and omissions are inevitable. The translators hope for the readers' understanding and assistance in addressing any shortcomings.

Translators, October 2023

Name Changes of Some Fungi in *Atlas of Chinese Macrofungal Resources* Mainly Since Publication

——Changes in Larger Ascomycetes

The species name in the 2015 Chinese edition of *Atlas of Chinese Macrofungal Resources*	The name of the fungi in the photo of the atlas which should be changed to	The publishing year of the changed name
Cordyceps gunnii (Berk.) Berk.	*Keithomyces neogunnii* (T.C. Wen & K.D. Hyde) Luangsa-ard et al.	2020
Cudonia sp. 1	*Cudonia claviformis* Zhu L. Yang	2023
Cudonia sp. 2	*Cudonia furfuracea* Zhu L. Yang	2023
Isaria cicadae Miq.	*Cordyceps chanhua* Z.Z. Li et al. 2021	2021

Name Changes of Some Fungi in *Atlas of Chinese Macrofungal Resources* Mainly Since Publication

——Changes in Jelly Fungi

The species name in the 2015 Chinese edition of *Atlas of Chinese Macrofungal Resources*	The name of the fungi in the photo of the atlas which should be changed to	The publishing year of the changed name
Auricularia delicata (Mont.) Henn.	*Auricularia sinodelicata* Y.C. Dai & F. Wu	2021
Auricularia mesenterica (Dicks.) Pers.	*Auricularia asiatica* Bandara & K.D. Hyde	2016
Exidia recisa (Ditmar) Fr.	*Exidia yadongensis* F. Wu et al.	2020
Tremella foliacea Pers.	*Phaeotremella yunnanensis* L.F. Fan et al.	2020

Name Changes of Some Fungi in *Atlas of Chinese Macrofungal Resources* Mainly Since Publication

——Changes in Coral Fungi

The species name in the 2015 Chinese edition of *Atlas of Chinese Macrofungal Resources*	The name of the fungi in the photo of the atlas which should be changed to	The publishing year of the changed name
Ramaria broomei (Cotton & Wakef.) R.H. Petersen	*Phaeoclavulina macrospora* Brinkmann	1897

Name Changes of Some Fungi in *Atlas of Chinese Macrofungal Resources* Mainly Since Publication

——Changes in Polyporoid, Hydnaceous & Thelephoroid Fungi

The species name in the 2015 Chinese edition of *Atlas of Chinese Macrofungal Resources*	The name of the fungi in the photo of the atlas which should be changed to	The publishing year of the changed name
Antrodia albida (Fr) Donk.	*Antrodia heteromorpha* (Fr.) Donk	1966
Antrodia carbonica (Overh.) Ryvarden & Gilb.	*Lentoporia subcarbonica* B.K. Cui et al.	2022

Antrodia hingganensis Y.C. Dai & Penttilä	*Adustoporia sinuosa* (Fr.) Audet	2017
Antrodia huangshanensis Y.C. Dai & B.K. Cui	*Cartilosoma ramentaceum* (Berk. & Broome) Teixeira	1986
Antrodia macrospora Bernicchia & De Domincis	*Dentiporus albidoides* (A. David & Dequatre) Audet	2017
Antrodia wangii Y.C. Dai & H.S. Yuan	*Fomitopsis bondartsevae* (Spirin) A.M.S. Soares & Gibertoni	2017
Auriscalpium vulgare Gray	*Auriscalpium microsporum* P.M. Wang & Zhu L. Yang (The fruiting bodies in the small round picture)	2019
Fomitopsis feei (Fr.) Kreisel	*Rhodofomes roseus* (Alb. & Schwein.) Kotl. & Pouzar	1990
Fomitopsis pinicola (Sw.) P. Karst.	*Fomitopsis subpinicola* B.K. Cui et al.	2021
Fomitopsis spraguei (Berk. & M.A. Curtis) Gilb. & Ryvarden	*Niveoporofomes orientalis* B.K. Cui & Shun Liu	2023
Hyphodontia radula (Pers.) Langer & Vesterh.	*Xylodon raduloides* Riebesehl & Langer	2017
Parmastomyces mollissimus (Maire) Pouzar	*Sarcoporia polyspora* P. Karst.	1894
Serpula similis (Berk. & Broome) Ginns	*Serpula dendrocalami* C.L. Zhao	2019

Name Changes of Some Fungi in *Atlas of Chinese Macrofungal Resources* Mainly Since Publication

——Changes in Cantharelloid Fungi

The species name in the 2015 Chinese edition of *Atlas of Chinese Macrofungal Resources*	The name of the fungi in the photo of the atlas which should be changed to	The publishing year of the changed name
Cantharellus cibarius Fr.	*Cantharellus applanatus* D. Kumari et al.	2013
Cantharellus minor Peck	*Cantharellus elongatipes* D. Kumari et al.	2013
Cantharellus tuberculosporus M. Zang	*Cantharellus versicolor* S.C. Shao & P.G. Liu	2016
Gomphus floccosus (Schwein.) Singer	*Turbinellus szechwanensis* (R.H. Petersen) Xue Ping Fan & Zhu L. Yang	2023
Gomphus fujishanensis (S. Imai) Parmasto	*Turbinellus parvisporus* Xue-Ping Fan & Zhu L. Yang	2023

Name Changes of Some Fungi in *Atlas of Chinese Macrofungal Resources* Mainly Since Publication

——Changes in Agarics

The species name in the 2015 Chinese edition of *Atlas of Chinese Macrofungal Resources*	The name of the fungi in the photo of the atlas which should be changed to	The publishing year of the changed name
Agaricus radicatus Schumach.	*Agaricus beijingensis* R.L. Zhao et al.	2021
Agaricus subrufescens Peck	Uncertain species of *Agaricus* sp.	–
Amanita hemibapha (Berk. & Broome) Sacc. s.l.	*Amanita rubroflava* Y.Y. Cui et al.	2018
Amanita manginiana Har. & Pat.	*Amanita caojizong* Zhu L. Yang et al.	2018
Crepidotus badiofloccosus S. Imai	*Crepidotus crocophyllus* (Berk.) Sacc.	1887
Crepidotus cinnabarinus Peck	*Crepidotus reticulatus* T. Bau & Y.P. Ge	2020
Gymnopus polyphyllus (Peck) Halling	*Gymnopus sinopolyphyllus* J.P. Li et al.	2022

Hygrocybe cuspidata (Peck) Murrill	*Hygrocybe rimosa* C.Q. Wang & T.H. Li	2021
Hygrocybe cf. *nigrescens* (Quél.) Kühner	*Hygrocybe rubroconica* C.Q. Wang & T.H. Li	2020
Hygrophorus cossus (Sowerby) Fr.	*Hygrophorus glutiniceps* C.Q. Wang & T.H. Li	2019
Lactarius camphoratus (Bull.) Fr.	*Lactarius rufus* (Scop.) Fr.	1838
Lactifluus hygrophoroides (Berk. & M.A. Curtis) Kuntz	*Lactifluus pseudohygrophoroides* H. Lee & Y.W. Lim	2017
Lactifluus subvellereus (Peck) Nuytinck	*Lactifluus pilosus* (Verbeken et al.) Verbeken	2011
Mycena capillaripes Peck	Uncertain species of *Mycena* sp.	–
Mycena galericulata (Scop.) Gray	*Mycena maculata* P. Karst.	1890
Mycena metata (Fr.) P. Kumm.	Uncertain species of *Marasmius* sp. or *Gymnopus* sp.	–
Mycena polygramma (Bull.) Gray	Uncertain species of *Mycena* sp.	–
Mycena sanguinolenta (Alb. & Schwein.) P. Kumm.	*Mycena haematopus* (Pers.) P. Kumm.	1871
Psathyrella gracilis (Fr.) Quél.	*Candolleomyces candolleanus* (Fr.) D. Wächt. & A. Melzer	2020
Psathyrella kauffmanii A.H. Sm.	*Psathyrella spintrigeroides* P.D. Orton	1960
Psathyrella multissima (S. Imai) Hongo	*Psathyrella boreifasciculata* Kytöv. & Liimat.	2014
Psathyrella olympiana A.H. Sm.	*Candolleomyces subminutisporus* T. Bau & J.Q. Yan	2021
Psathyrella spadiceogrisea (Schaeff.) Maire	*Candolleomyces subcacao* T. Bau & J.Q. Yan	2021
Psathyrella spintrigeroides P.D. Orton	*Psathyrella phegophila* Romagn.	1985
Russula cf. *amoena* Quél.	*Russula pseudoamoenicolor* A. Ghosh et al.	2016
Russula japonica Hongo	*Russula brevispora* Y.L. Chen & J.F. Liang	2023
Squamanita umbonata (Sumst.) Bas	*Squamanita sororcula* J. W. Liu & Zhu L. Yang	2021
Strobilurus sp.	*Strobilurus orientalis* Zhu L. Yang & J. Qin	2018
Tricholoma pardinum (Pers.) Quél.	*Tricholoma sinopardinum* Zhu L. Yang et al.	2017

Name Changes of Some Fungi in *Atlas of Chinese Macrofungal Resources* Mainly Since Publication

——Changes in Boletes

The species name in the 2015 Chinese edition of *Atlas of Chinese Macrofungal Resources*	The name of the fungi in the photo of the atlas which should be changed to	The publishing year of the changed name
Boletellus chrysenteroides (Snell) Snell	*Boletellus putuoensis* N.K. Zeng et al.	2022
Boletellus emodensis (Berk.) Singer	*Boletellus areolatus* Hirot. Sato	2015
Boletellus longicollis (Ces.) Pegler & T.W.K. Young	*Aureoboletus viscosus* (C.S. Bi & Loh) G. Wu & Zhu L. Yang	2016
Boletellus obscurecoccineus (Höhn.) Singer	*Boletellus puniceus* (W.F. Chiu) X.H. Wang & P.G. Liu	2002
Boletinellus merulioides (Schwein.) Murrill	*Psiloboletinus lariceti* (Singer) Singer	1945
Boletus bicolor Raddi	*Lanmaoa asiatica* G. Wu & Zhu L. Yang	2015

Boletus hortonii A.H. Sm. & Thiers	*Hemileccinum* cf. *rugosum* G. Wu & Zhu L. Yang	2016
Boletus inedulis (Murrill) Murrill	*Butyriboletus hainanensis* N.K. Zeng et al.	2016
Boletus kermeinus Har. Takah. et al.	*Butyriboletus ruber* (M. Zang) Kui Wu et al.	2020
Boletus meiweiniuganjun Dentinger	*Boletus bainiugan* Dentinger	2013
Boletus nigropunctatus W.F. Chiu	*Hourangia cheoi* (W.F. Chiu) Xue T. Zhu & Zhu L. Yang	2015
Boletus paluster Peck	Uncertain species of *Suillus* sp.	–
Boletus quercinus Hongo	*Caloboletus guanyui* N.K. Zeng et al.	2019
Boletus subtomentosus L.	*Xerocomus* cf. *rugosellus* (W.F. Chiu) F.L. Tai	1979
Boletus subvelutipes Peck	*Lanmaoa* cf. *angustispora* G. Wu & Zhu L. Yang	2015
Boletus umbriniporus Hongo	*Neoboletus* cf. *brunneissimus* (W.F. Chiu) Gelardi et al.	2014
Gyroporus cyanescens (Bull.) Quél.	*Gyroporus alpinus* Yan C. Li et al.	2021
Heimioporus retisporus (Pat. & C.F. Baker) E. Horak	*Heimioporus gaojiaocong* N.K. Zeng & Zhu L. Yang [with a fruitbody of *Heimioporus japonicus* (Hongo) E. Horak at the last in the photo]	2018
Mucilopilus castaneiceps (Hongo) Hid. Takah.	*Mucilopilus paracastaneiceps* Yan C. Li & Zhu L. Yang	2021
Pulveroboletus auriflammeus (Berk. & M.A. Curtis) Singer	*Aureoboletus miniatoaurantiacus* (C.S. Bi & Loh) Ming Zhang et al.	2019
Retiboletus griseus (Frost) Manfr. Binder & Bresinsky	*Retiboletus fuscus* (Hongo) N.K. Zeng & Zhu L. Yang	2016
Retiboletus nigerrimus (R. Heim) Manfr. Binder & Bresinsky	*Retiboletus* cf. *zhangfeii* N.K. Zeng & Zhu L. Yang	2016
Rubinoboletus balloui (Peck) Heinem. & Rammeloo	*Tylopilus rubrotinctus* Yan C. Li & Zhu L. Yang	2021
Sinoboletus gelatinosus M. Zang & R.H. Petersen	*Aureoboletus duplicatoporus* (M. Zang) G. Wu & Zhu L. Yang	2016
Suillus bovinus (L.) Roussel	*Suillus* cf. *pinetorum* (W.F. Chiu) H. Engel & Klofac	1996
Xanthoconium cf. *affine* (Peck) Singer	*Xanthoconium sinense* G. Wu et al.	2016
Xerocomellus rubellus (Krombh.) Šutara	*Hortiboletus subpaludosus* (W.F. Chiu) Xue T. Zhu & Zhu L. Yang	2016

Name Changes of Some Fungi in *Atlas of Chinese Macrofungal Resources* Mainly Since Publication

——Changes in Gasteroid Fungi

The species name in the 2015 Chinese edition of *Atlas of Chinese Macrofungal Resources*	The name of the fungi in the photo of the atlas which should be changed to	The publishing year of the changed name
Dictyophora duplicata (Bosc) E. Fisch.	*Phallus ultraduplicatus* X.D. Yu et al.	2015
Dictyophora indusiata (Vent.) Desv.	*Phallus rigidiindusiatus* T. Li et al.	2021
Dictyophora multicolor Berk. & Broome	*Phallus luteus* (Liou & L. Hwang) T. Kasuya	2009
Phallus impudicus L.	*Phallus dongsun* T.H. Li et al.	2020
Phallus rubicundus (Bosc) Fr.	*Phallus rugulosus* (E. Fisch.) Lloyd , Yu Li, Taihui Li, Zhuliang Yang, Tolgor Bau, & Yucheng Dai	1908

Contents

Editorial Committee

Authors

Yu Li, Taihui Li, Zhuliang Yang, Tolgor Bau, Yucheng Dai

Translators

Yu Li, Taihui Li, Bo Zhang, Liangcheng Chen, Pu Liu, Ming Zhang, Ting Li, Jipeng Li, Jiajun Hu, Yang Wang, Jiahui Xing

Cantharelloid Fungi: Species with Names Beginning with C-G

902 *Cantharellus cibarius* Fk

Basidiomata 4–12 cm high, fleshy, tubiform or trumpet-shaped, vivid apricot yellow to yolk yellow. Pileus 3–12 cm in diam., applanate at first, then depressed at centre, glabrous, subviscid when young; margin wavy, sometimes valvate, incurved. Context 2–4 mm thick, white to yolk yellow, with an almond flavour. False lamellae decurrent, ridged-lamellate, narrow and distant, furcate or interweaved. Stipe 2–8 cm long, 0.8–2 cm thick, tapering downwards, apricot yellow, glabrous, solid. Basidiospores 7–10 × 5–6.5 pm, elliptical, smooth, hyaline.

In summer to autumn, on the ground in coniferous or mixed coniferous and broadleaf forests. Edible. Known from most parts of China.

903 *Cantharellus minor* Peck

Pileus 1–3 cm in diam., subhemispherical to applanate when young, depressed at disc, shallowly valvate to trumpet-shaped, margin incurved or irregularly wavy, involute, apricot to bright yellow, yolk yellow, egg yellow to orange yellow, glabrous. Context thin, brittle, pale yellow to pale orange yellow, with a light fruity aroma. False lamellae decurrent to subdecurrent, distant, narrow ridged, egg yellow to orange yellow. Stipe 1.5–4 cm long, 0.2–0.7 cm thick, cylindrical, equal or tapering downwards, concolorous with pileus or paler, solid, becoming hollow at maturity. Basidiospores 6–10 × 4.5–6 im, elliptical to oval, smooth, pale yellow to pale ochraceous, apiculate.

© Central China Farmer's Publishing House 2025
Y. Li et al., *Atlas of Chinese Macrofungal Resources*, https://doi.org/10.1007/978-981-96-0502-6_1

In summer to autumn, gregarious under mixed coniferous and broadleaf forests. Edible. Known from most parts of China.

904 *Cantharellus tuberculosporus* M. Zang

Pileus 3–7 cm in diam., applanate, depressed at centre, yellow to yellow brown, smooth or with lacerated epidermis into yellow brown scales, yellow to bright yellow at margin. Context thin. False lamellae not typically lamellate, folded to ridged, yellow to pale yellow. Stipe 3–7 cm long, 0.5–1 cm thick, cylindrical, yellow to pale yellow, solid. Basidiospores 7.5–9.5 × 5.5–6.5 pm, elliptical, smooth.

In summer and autumn, on the ground in subalpine mixed coniferous and broadleaf forests. Edible, distributed in the Qinghai-Tibet region.

905 *Craterellus aureus* Berk. & M.A. Curtis

Basidiomata 7–12 cm high, golden yellow to old golden yellow, subtubiform or nearly trumpet-shaped. Pileus 2–6 cm in diam., irregularly wavy and incurved or reflexed at margin, subglabrous, waxy, deeply concave at centre to stipe, without an obvious boundary with stipe. False lamellae absent. Hymenophore even to nearly so. Stipe and pileus confluent to trumpet-shaped, 2–6 cm long, 3–8 mm wide, tapering downward base. Basidiospores 7.5–10 × 6–7.5 pm, elliptical, smooth, hyaline.

In summer, gregarious or caespitose on ground in broadleaf forest with Fagaraceae trees. Edible. Distributed in southern China.

906 *Craterellus cornucopioides* (L.) Pers.

Pileus 3–8 cm in diam., deeply concave at centre, grey, greyish brown to greyish black, squamulose, wavy or incurved at margin. Context thin. False lamellae absent. Hymenophore pale grey to greyish purple, even to nearly so. Stipe 2–5 cm long, 5–10 mm wide, tapering downward, grey, greyish brown to greyish black, hollow. Basidia mainly 2-spored. Basidiospores 10–15 × 6–10 gm, elliptical, smooth.

In summer to autumn, on the ground in coniferous forests or mixed coniferous and broadleaf forests. Edible. Distributed in most parts of China.

907 *Craterellus lutescens* (Fr.) Fr.

Pileus 3–8 cm in diam., trumpet-shaped, yellow brown to greyish yellow, squamulose, wavy or incurved at margin. Context thin. False lamellae atypical, subeven to veined, decurrent, yellow to yellow brown. Stipe 4–8 cm long, 5–10 mm wide, cylindrical, yellow, golden yellow to orange yellow, hollow. Basidiospores 9–12 × 6–8.5 gm, elliptical, smooth.

In summer to autumn, on the ground in coniferous forests or mixed coniferous and broadleaf forests. Edible. Distributed in most parts of China

908 *Gloeocantharellus persicinus* T.H. Li et al.

Pileus slightly asymmetrical, 3.8–7 cm in long diameter, 3.5–6 cm in short diameter, convex to applanate, or shallowly infundibuliform, peach colour to pale orange red; margin wavy or valvate, involuted at first. Context white, unchanging when bruised, yellow brown when dried. False lamellae decurrent, crowded, white to cream with light greenish grey tinct, smoky grey or greenish brown when dried, unequal, furcate. Stipe 4–4.5 cm long, 9–11 mm thick, cylindrical, central to eccentric, light peach colour, light yellow to pinkish yellow, concolorous with pileus or paler on the places with red tinge. Basidiospores 4.5–5 × 7–12 gm, long ellipsoid, coarsely verruculose, brown to yellow brown. Gloeoplerous hyphae 5–7 × 40–50 gm, immersed to slightly projecting beyond hymenium, brown to yellow brown.

In summer to autumn, on the ground in the broadleaf forest. Distributed in southern China.

909 *Gomphus floccosus* (Schwein.) Singer

Pileus 3–7 cm in diam., trumpet-shaped, yellow to orange red, with red scales, deeply concave centrally to the base of stipe. False lamellae atypical or absent, folded, decurrent, dirty white to pale yellow. Stipe 3–7 cm long, 5–15 mm thick, cylindrical, dirty white to pale yellow. Basidiospores 11–15 × 6–7.5 pm, elliptical, smooth to slightly rough. Clamp connection absent.

In summer to autumn, on the ground under coniferous forests. Reported to be poisonous. Not recommended to be eaten. Distributed in most parts of China.

910 *Gomphus fujisanensis* (S. Imai) Parmasto

Pileus 5–8 cm in diam., trumpet-shaped, pale pinkish brown to pale yellow brown streaked with yellowish white, with pale greyish brown scales, centrally concave to the base of the stipe. False lamellae atypical, folded, decurrent, dirty white, beige to light brown. Stipe 3–8 cm long, 5–20 mm thick, dirty white. Basidia 60–80 × 10–12 pm. Basidiospores 14–18 × 6–7.5 pm, elliptical, slightly rough.

In summer to autumn, on the ground under coniferous forests. Edible. Distributed in most parts of China.

911 *Gomphus kauffmanii* (A.H. Sm.) Corner

Pileus 2–15 cm in diam., deeply concave at centre to trumpet shaped; margin thin, wavy, often lacerated when mature; surface cream to tawny or pale pinkish cinnamon, with erect or upturned dark brown scales. Context thin, fibrous, white. False lamellae decurrent, obtusely ridged longitudinally, intervenose, forked, cream to light brown, darker when mature. Stipe 6–12 cm long, 2–4 cm thick, tapering downward, solid at first, then gradually hollow, cream when young, pale brown to dark brown when old, changing to pale pinkish purple when bruised, smooth. Basidiospores 12–18 × 6–8 pm, elliptical, slightly rough.

In summer to autumn, solitary, scattered to gregarious on the ground in coniferous forests. Edible. Distributed in northeast China.

912 *Gomphus orientalis* R.H. Petersen & M. Zang

Basidiomata often branched at base. Pileus 3–10 cm in diam., applanate, slightly depressed at centre, light brown to purplish, squamulose. False lamellae decurrent, pale brown to pale purple. Stipe 1–3 cm long, 1–2 cm thick, greyish brown to purplish. Basidia 80–100 × 10–13 gm. Basidiospores 10–16 × 4.5–7.5 gm, elliptical, verruculose.

In summer to autumn, on the ground in subalpine forests. Edible. Distributed in Qinghai-Tibet region.

Agarics 1: Species with Names Beginning with A-B

913 *Agaricus abruptibulbus* Peck

Pileus 4–10 cm broad, convex to oblate hemispherical, obtusely umbonate at centre, plane in later stage; white or yellowish white, darker at disc, with velar remnants at margin. Context thick, white or yellowish. Lamellae free, greyish white when young, becoming light yellowish brown, eventually dark purplish brown. Stipe 5–15 cm long, 1–3 cm thick, cylindrical, bulbously enlarged at base. Annulus superior, white, membranous, fugacious. Basidiospores 6–9 × 4–5 μm, elliptical to broadly elliptical, smooth, dark yellowish brown to dark brown.

In summer to autumn, gregarious or scattered on ground in mixed forests or grassland in margin of forests. Edible. Distributed in Central and Southern China.

This species is similar to *A. silvicola*, but stipe of *A. silvicola* is not bulbous at base and the basidiospores are smaller.

914 *Agaricus arvensis* Schaeff.

Pileus 5.5–14 cm broad, hemispherical at first, then oblate hemispherical to convex, subplane when mature, sometimes with a rounded prominence at disc, nearly white when fresh, dirty white at centre, then becoming light yellow to ochraceous, glabrous; often lacerated at margin. Context white, thick. Lamellae free, pink at first, brown to dark brown when mature, close, unequal. Stipe 4–10.5 cm long, 1.5–2.5 cm thick, central, subcylindrical, concolourous with pileus, solid at first, becoming hollow, unchanging when injured, sometimes enlarged at base. Annulus superior, membranous, large and thick, white, fugacious. Basidiospores 7.3–9.7 × 4.8–6.1 μm, elliptical to oval, smooth, yellowish brown to dark brown.

Solitary on grassland. Edible. Distributed in Northeast, Inner Mongolia, and Northwest China.

© Central China Farmer's Publishing House 2025
Y. Li et al., *Atlas of Chinese Macrofungal Resources*, https://doi.org/10.1007/978-981-96-0502-6_2

915 *Agaricus augustus* Fr.

Pileus 9–20 cm broad, subglobose at first, becoming flat hemispherical, then plane, densely with brown scales; scales massive at disc. Context thick, white, becoming yellow when injured. Lamellae free, greyish white when young, becoming pink and eventually dark chocolate brown to blackish. Stipe 8–17 cm long, 2–3.5 cm thick, enlarged at base, glabrous above annulus, squamulose below annulus. Annulus double, superior, membranous, white or hay yellow. Basidiospores 7–9.5 × 5–6.5 μm, elliptical to suboval, smooth, brown.

In summer to autumn, gregarious on ground in mixed forests. Edible. Distributed in Qinghai Tibet Region, Northeast, and Northwest China.

This species is similar to *A. subrufescens*, whereas the basidiospores of the latter species are smaller.

916 *Agaricus bisporus*(J.E. Lange) Imbach

Pileus 4–10 cm broad, subhemispherical to convex; often incurved at margin; nearly white to brown, becoming light yellow to yellowish brown; appressed-fibrillose or minutely squamulose, roughly cracked in dry conditions. Context white, becoming light red when injured. Lamellae pinkish to pink at first, becoming dark brown to nearly black, free, unequal. Stipe 3–6 cm long, 1–2 cm thick, subcylindrical, white, soft or solid insides. Annulus single, superior to medial, membranous, white, fagacious. Basidiospores 5–8 × 4–6.5 μm, elliptical, smooth, brown.

Scattered to subgregarious on alpine grassland, lawn, forest, cultivation, park and roadside. A famous edible fungus, with white and brown varieties, extensively cultivated in China. Distributed in Qinghai Tibet Region, Northeast, and Central China.

917 *Agaricus blazei* Murrill

Pileus 5.5–10.5 cm broad, hemispherical to oblate hemispherical at first, becoming plane when mature, sometimes plane or concave at centre, with light brown or light yellowish brown fibrillose scales, with velar remnants at margin. Context white, thick, becoming yellow when injured. Lamellae free, white to pink at first, brown to blackish brown when mature, close, fragile. Stipe 5.5–11.5 cm long, 1.2–2.2 cm thick, subequal or slightly enlarged at base, sometimes slightly crooked, squamulose below annulus when young, solid. Annulus superior, white, membranous, thin, with brown cottony materials underneath, fagacious. Basidiospores 5.3–6.6 × 3.7–4.9 μm, broadly elliptical to oval, smooth, light brown to brown.

In summer to autumn, gregarious on grassland. Edible and medicinal. Cultivated widely in China. Distributed in Qinghai Tibet Region, Northeast, and Southern China.

918 *Agaricus campestris* L.

Pileus 5–10 cm broad, oblate hemispherical at first, then becoming plane, sometimes depressed at centre, white to creamy white, glabrous, or with flocculose scales in later stage, often split at margin in dry conditions. Context white. Lamellae free, pink at first, becoming brown and then dark chocolate brown at maturity. Stipe 2–10 cm long, 1–2 cm thick, subglabrous or slightly fibrillose, white. Annulus single, superior, membranous, white, fagacious. Basidiospores 7–9 × 4.5–6 μm, elliptical, greyish brown to brown, smooth.

In spring to autumn, solitary or gregarious on grasslands, roadside, fields, compost sites, and open fields in forests. Edible. Distributed widely in China.

This species is similar to *A. arvensis*, but the latter species is becoming yellow when injured and has an enlarged stipe base.

919 *Agaricus crocopeplus* Berk. & Broome

Pileus 3–6 cm broad, globose to oblate hemispherical at first, becoming plane when mature, with orange long hairs or flocculose scales, with velar remnants at margin. Context white or dirty white, becoming light brown in later stage, 3–4 mm thick at stipe. Lamellae free, close, unequal, dirty white to light brown at first, becoming brown at maturity. Stipe 3–6 cm long, 5–8 cm thick, cylindrical, hollow when mature, with long hairs concolourous with pileus. Annulus superior, atypical, as remnants of universal veil, homogeneous with scales. Basidiospores 5–8 × 3.5–4.5 μm, elliptical to oval, smooth, greyish brown.

In summer to autumn, on ground in forests. Distributed in Central and Southern China.

This species is similar to *Cystoagaricus trisulphuratus* and *Hypholoma cinnabarinum*, but their relationships still need to be studied.

920 *Agaricus dulcidulus* Schulzer

Pileus 2–7 cm broad, convex at first, often incurved at margin, becoming plane, with dense purplish brown to light pink fibrils on greyish pink background, darker at disc. Context white, becoming yellow when injured. Lamellae free, close, light greyish brown at first, dark purplish brown in later stage. Stipe 2–5 cm long, 4–8 mm thick, cylindrical, enlarged to clavate to bulbous at base; superior white, becoming darker from apex to base, light brown. Annulus superior to medial, nearly white, membranous, fagacious. Basidiospores 4.5–6 × 3.5–4 μm, elliptical, smooth.

In summer to autumn, scattered on ground in broad-leaved forests. Edible. Distributed in Southern China.

921 *Agaricus moelleri* Wasser

Pileus 6–7 cm broad, broadly convex to plane, broadly umbonate at disc, dirty white, becoming pinkish when mature, covered with grey or dark grey scales, blackish at disc. Context white. Lamellae free, pink at first, then becoming to pinkish brown. Stipe 6–7 cm long, 5–8 mm thick, cylindrical, marginately bulbous at base, white, yellow inside. Annulus superior to medial, membranous, larger, dirty white, becoming yellow when injured. Basidia 18–20 × 6–7 μm. Basidiospores 4.5–5.5 × 3–3.5 μm, elliptical, smooth, brown.

In summer to autumn, on ground in broad-leaved forests. Distributed widely in China.

922 *Agaricus porphyrizon* P.D. Orton

Pileus 5–8 cm broad, hemispherical at first, then convex to plane, sometimes concave at centre, covered with dark red to pinkish brown and persistent scales, becoming paler when mature, often incurved and lacerated at margin. Context white, thick, with an almond flavor. Lamellae free, white when young, grey to purplish black when mature, serrulated at edge, close. Stipe 5–8 cm long, 5–10 mm thick, white, cylindrical, gradually enlarged toward base, becoming hollow when mature. Annulus superior, membranous, white. Basidiospores 5–6.5 × 3.5–4.5 μm, elliptical, smooth, brown to brown purple.

In summer to autumn, solitary or gregarious on ground in forests or parks. Poisonous. Distributed in Northeast and Northern China.

923 *Agaricus radicatus* Schumach.

Pileus 4–9 cm broad, dirty white, subglobose at first, then becoming plane, covered with brown or light brown scales at disc, more sparsely scaly at margin. Context white, thick, slightly becoming dark red when injured. Lamellae free, greyish white to pink at first, becoming brown to black brown, close. Stipe 5–6.5 cm long, 8–12 mm thick, white; scales below annulus white, becoming brown gradually, fagacious in later stage; base enlarged, short radicant, becoming light yellow when injured. Annulus single, superior, membranous, white, fagacious. Basidiospores 6.5–8 × 4.5–5.5 μm, elliptical, smooth, brown.

In autumn, solitary or scattered on ground in forests. Edible. Distributed in Northern China.

924 *Agaricus silvaticus* Schaeff

Pileus 5–10 cm broad, oblate hemispherical, then becoming plane gradually, nearly white, covered with light brown or reddish brown scales at centre, with sparser scales at margin, radially split in dry conditions at margin. Context white. Lamellae free, white at first, becoming pink gradually, then chestnut brown to black brown with age. Stipe 6–10 cm long, 1–1.5 cm thick, white, with white fibrillose scales above annulus, slightly enlarged at base, becoming dirty yellow when injured. Annulus single, superior, membranous, white. Basidiospores 5–6.5 × 3.5–4.5 μm, elliptical, smooth, light brown.

In summer to autumn, solitary to gregarious on ground in mixed forests. Edible. Distributed in Northeast, Northern, Northwest and Central China.

925 *Agaricus silvicola* (Vittad.) Peck

Pileus 5–11 cm broad, subglobose to oblate hemispherical at first, becoming convex, plane at maturity, white or light yellow, light brown at centre, with appressed silky fibrils, often split at margin. Context white. Lamellae free, white at first, becoming pink, brown, or black brown. Stipe 6–13 cm long, 0.5–1.5 cm thick, dirty white, subcylindrical, slightly enlarged at base, becoming yellow when injured. Annulus single, superior, membranous, white, pendant. Basidiospores 5–6.5 × 3–4.5 μm, globose to oval, smooth, dark brown.

In summer to autumn, solitary or scattered on ground in forests. Edible. Distributed in Northeast, Northern, Northwest and Central China.

926 *Agaricus subrufescens* Peck

Pileus 5–15 cm broad, oblate hemispherical at first, becoming convex, subplane at maturity; white, greyish white to pinkish brown; densely with retrocurling flocculent scales, split at margin. Context thin, white. Lamellae free, white at first, becoming pink, blackish brown at maturity. Stipe 6–15 cm long, 1–1.5 cm thick, concolourous with pileus and scaly below annulus. Annulus double, superior, membranous, nearly white. Basidiospores 6–7.5 × 4–5 µm, elliptical, smooth, dark purplish brown.

In autumn, solitary, gregarious to subcaepitose on ground in forests. Edible. Distributed in Northeast, Northern, Northwest and Central China.

927 *Agaricus urinascens* (Jul. Schäff. & F.H. Møller) Singer

Pileus 9–16 cm broad, globose to oblate hemispherical at first, gradually plane at maturity; nearly white at first, gradually becoming light yellow at maturity; covered with punctate appressed brown scales. Context thick, white. Lamellae free, nearly white, becoming pink to black brown, close, unequal. Stipe 5.5–8 cm long, 1.2–1.8 cm thick, white, with light yellow scales, soft to solid inside, slightly enlarged at base, tapering upwards. Annulus single, medial to superior, white, membranous, larger and thick, persistent. Basidiospores 10.5–12 × 6–7 µm, elliptical, smooth, brown.

In spring to autumn, solitary to gregarious on grassland. Edible. Distributed in Northeast, Northern, Northwest and Central China.

928 *Agaricus xanthodermus* Genev.

Pileus 4–8 cm broad, convex to subspherical at first, gradually plane at maturity, dirty white, brownish at disc, glabrous; margin incurved, light yellow. Context white. Lamellae free, light pink to black brown, close. Stipe 5–15 cm long, 1–2 cm thick, cylindrical, enlarged at base, white, glabrous, solid when young, becoming hollow when mature, enlarged at base, with a yellow bulbous base. Annulus superior to medial, membranous. Basidiospores 5–6.5 × 3–4.5 µm, elliptical, smooth, brown.

In summer to autumn, solitary on ground in forests. Poisonous. Distributed in Qinghai-Tibet Region and Northwest China.

929 *Agrocybe* cf. *broadwayi* (Murrill) Dennis

Pileus 2–4 cm broad, hemispherical, convex to subplane, yellowish white, yellowish brown to light orange yellow brown, sometimes greyish at margin, slightly slimy to slimy, sometimes reticulately rugulose, sometimes split at margin when old. Context thick, white to nearly white, mild. Lamellae adnate or sinuate, close, unequal, separable from stipe at maturity, smoky brown, brown, to dark brown. Stipe 4–6.5 cm long, 3–6 mm thick, cylindrical, slightly enlarged at base, yellowish white, light brown, striate, with white mycelium at base. Annulus fagacious. Basidiospores 12.5–15 × 6.5–10 µm, elliptical, smooth, with a truncated apical germ-pore, dark brown.

In summer to autumn, scattered on ground in broad-leaved forests. Distributed in Southern China.

930 *Agrocybe cylindracea* (DC.) Maire

Pileus 3–10 cm broad, hemispherical at first, then convex to subplane, dark brown to tawny when young, becoming light brown to light yellow, lighter at margin, darker at centre, subslimy when wet, glabrous or rugulose; margin with a veil and incurved when young. Context white or dirty white. Lamellae adnate to decurrent, close, unequal, white at first, brown with age. Stipe 3–13 cm long, 0.5–1.5 cm thick, cylindrical, subequal, dirty white, dirty brown at base, fibrillose. Annulus superior, membranous, white at first, with dark brown basidiospores on surface with age, persistent. Basidiospores 7–10 × 4–6 µm, elliptical, smooth, dark brown.

In spring, caespitose or solitary on dead wood of *Populus* and *Salix*. Cultivated widely in China, product name is "agrocybe". Edible. Distributed in Southern and Central China.

931 *Agrocybe erebia* (Fr.) Kühner ex Singer

Pileus 1.5–7 cm broad, truncated-conical at first, becoming hemispherical gradually, plane at maturity, obtusely umbonate at centre; incurved at margin when young, upturned to reflexed at margin at maturity, glabrous, slimy, dark brown and remaining fairly dark, eventually dull brown, often with whitish partial velar remnants at margin. Context thin, dirty white. Lamellae adnate to decurrent, close, unequal, light brown to greyish brown at first. Stipe 2.5–6.5 cm long, 0.3–1 cm thick, cylindrical, often slightly enlarged at base, solid, dirty white to light brown, darker at base. Annulus superior to medial, membranous, white, persistent. Basidiospores 7.8–13.6 × 4.8–7.8 μm, elliptical-oblong to oval, narrower at apex, smooth, without a apical germ-pore, dark brown.

In spring to autumn, gregarious on ground in broad-leaved forests. Edible. Distributed in Northeast and Northwest China.

932 *Agrocybe pediades* (Fr.) Fayod

Pileus 1–3 cm broad, hemispherical when young, then plano-umbonate broadly; dark brown to light yellow, glabrous, slimy when wet; incurved at margin when young, becoming plane at maturity. Context white to light yellow, very thin, unchanging when injured. Lamellae adnate, milky white at first, becoming brown to rusty at maturity, close, unequal. Stipe 2–7 cm long, 1–2 mm thick, subcylindrical, central, often twisted, concolourous with pileus, fibrillose, solid at first, becoming hollow at maturity. Annulus fibrillose, fagacious. Basidiospores 11–14 × 7–8 μm, elliptical, smooth, dark brown.

Scattered or gregarious on grassland. Edible, but easily confused with some poisonous mushrooms. Distributed in Southern China.

933 *Agrocybe praecox* (Pers.) Fayod

Pileus 2–8 cm broad, conical at first, then oblate hemispherical to plane with a prominent disc, becoming plane at maturity, sometimes slightly convex, hygrophanous ochraceous to light brown or light greyish brown when wet; margin incurved when young, then plane, sometimes white, slimy when wet, glabrous, rugulose or areolate, with partial velar remnants when young. Context white to light yellow, very thin. Lamellae adnate or subsinuate, close, unequal, light brown at first, becoming dark brown, concolourous to slightly darker and serrulate at edge. Stipe 3–10 cm long, 0.3–1.2 cm thick, white to light brown, slightly enlarged and with white rhizomorphs at base. Annulus superior, membranous, white, fagacious. Basidiospores 8–13 × 6.5–8 μm, oval to elliptical, with an obvious apical germ-pore, smooth, melline yellow.

In spring, scattered or gregarious on ground in forests or fields, grassland of roadside. Edible. Distributed widely in China.

934 *Agrocybe salicacicola* Zhu L. Yang et al.

Pileus 4–8 cm broad, hemispherical at first, becoming oblate hemispherical gradually, plane at maturity, sometimes slightly concave at centre, incurved at margin when young, expanded at maturity, beige at disc when mature, becoming lighter to white toward margin; glabrous or areolate at disc, not slimy, estriate; often with velar remnants at margin. Context white, thin, unchanging when injured. Lamellae decurrent, close, light brown at first, greyish brown when mature. Stipe 5–10 cm long, 5–8 mm thick, subcylindrical, dirty white, with powdery brown squamule, tapering downwards. Annulus larger, membranous, white, persistent to fagacious. Basidiospores 6.3–10.3 × 4.4–6.3 μm, elliptical, smooth, dark brown.

In spring, clustered on dead wood. Edible, cultivated. Distributed in Central China.

This species is similar to *A. cylindracea*, but the latter species has darker pileus and adnate lamellae.

935 *Amanita altipes* Zhu L. Yang et al.

Pileus 4–9 cm broad, oblate hemispherical to plane, light yellow to yellow, darker at centre, with light yellow, yellow to dirty yellow flocculent velar remnants, striate-ribbed at margin. Context white. Lamellae white to light yellow, light yellow to yellow at edge. Stipe 9–16 cm long, 0.5–1.8 cm thick, light yellow; bulbous to suboval at enlarged base, with yellow to light yellow velar remnants at upper portion. Annulus superior. Basidiospores 8–10 × 7.5–9.5 μm, globose to subglobose, smooth, hyaline, inamyloid.

In summer to autumn, on ground in sub-alpine coniferous, broad-leaved and mixed forests. Might poisonous. Distributed in Qinghai-Tibet Region.

936 *Amanita castanopsidis* Hongo

Pileus 6–13 cm broad, hemispherical to oblate hemispherical at first, plane at maturity, with conical or pyramidal velar remnants, white at first, then becoming dirty white, grey to brown at apex when dried, with smaller velar remnants at margin; margin flocculent and even. Context white, without special odour. Lamellae free to subfree, white. Stipe 6–13 cm long, 1–2 cm thick, subcylindrical, white, solid; bas enlarged, radicant. Annulus easily cracked and fagacious. Basidiospores 9.3–12.6 × 7.5–9.1 μm, elliptical to elliptical-oblong, smooth.

In summer to autumn, on ground in broad-leaved forests. Distributed in Central and Southern China.

937 *Amanita ceciliae* (Berk. & Broome) Bas s.l.

Pileus 6–15 cm broad, campanulate at first, becoming hemispherical gradually, then plane; dark yellow to greyish brown at first, then becoming darker; with greyish black pulverulent granules at centre, striate at margin. Context thin, white. Lamellae white or light grey, close, free, unequal. Stipe 12–19 cm long, 1–2 cm thick, cylindrical, slightly enlarged at base, white at upper portion, light grey and with white hairy scales at lower portion. Volva as 2–3 pulverulent rings. Basidiospores 11–13.5 × 10–13.5 μm, subglobose, smooth, hyaline.

In summer to autumn, scattered on ground in forests. Edible. Distributed in Northeast China.

938 *Amanita citrina* (Schaeff.) Pers.

Pileus 5–8 cm broad, light yellow to yellow, darker at disc, with light yellow massive scales, even at margin. Context white. Stipe 6–10 cm long, 0.8–1.5 cm thick, white to light yellow, truncated at base. Annulus superior, membranous, white to light yellow. Basidiospores 7–9.5 × 7–9 μm, globose to subglobose, smooth, hyaline, amyloid.

In summer to autumn, on ground of Fagaceae and Pinaceae forests. Poisonous. Distributed widely in China.

939 *Amanita clarisquamosa* (S. Imai) S. Imai

Pileus 4–10 cm broad, dirty white, brownish at disc, with greyish brown to brown, ragged or membranous to fibrillous velar remnants, shortly ribbed at margin. Context white. Stipe 6–13 cm long, 1–2 cm thick, white. Volva saccate, white to dirty white. Annulus superior, easily cracked, with greyish brown pulverulent scales. Basidiospores 10–13.5 × 6–7 μm, elliptical to elliptical-oblong, smooth, hyaline, amyloid.

In summer to autumn, on ground of Fagaceae and Pinaceae forests. Poisonous. Distributed widely in China.

940 *Amanita exitialis* Zhu L. Yang & T.H. Li

Pileus 4–8 cm broad, white, sometimes beige at disc, even at margin. Stipe 7–9 cm long, 0.5–1.5 cm thick, white; bulbous at base, 1–3 cm thick; volva shallowly. Annulus superior to subsuperior, membranous. Yellow in KOH. Basidia 2-sterigmate. Basidiospores 9.5–12 × 9–11.5 μm, globose to subglobose, smooth, hyaline, amyloid.

In spring to early summer, on ground of *Castanopsis fissa* (Figure) forest. Highly poisonous, it should be pay more attention to discriminate this species when collecting wild white mushroom. Distributed in Central and Southern China.

941 *Amanita farinosa* Schwein

Pileus 3–5 cm broad, light grey to light brown; long ribbed at margin; velar remnants pulverulent, sometimes warty or flocculent, grey to brownish grey. Context white. Lamellae white, close. Stipe 5–8 cm long, 3–6 mm thick, subcylindrical or tapered upwards, white, subbulbous to oval at base, covered with grey to brownish grey pulverulent velar remnants at upper portion. Annulus absent. Basidiospores 6.5–8 × 5.5–7 µm, subglobose to broadly elliptical, smooth, hyaline, inamyloid.

In summer to autumn, on ground of forests. Poisonous. Distributed widely in China.

942 *Amanita flavipes* S. Imai

Pileus 3.5–12 cm broad, light yellow to yellowish brown; velar remnants on surface yellow to light yellow, granular to verrucose. Stipe 5–15 cm long, 0.5–2 cm thick, white, light yellow, yellow; subbulbose at base, oval or ventricose, 1.5–4 cm thick; covered with light yellow to yellow, pulverulent to verrucose velar remnants at upper portion. Annulus superior. Basidiospores 7–9 × 5.5–7 µm, broadly elliptical to elliptical, smooth, hyaline, amyloid.

In summer to autumn, on ground of coniferous and mixed forests. Suspected poisonous, not edible. Distributed widely in China.

943 *Amanita fritillaria* Sacc.

Pileus 4–10 cm broad, light grey, brownish grey to light brown, with looming radiate **fibrillous** patterns, with dark grey to nearly black scales. Stipe 5–10 cm long, 0.6–1.5 cm thick, white to dirty white, covered with grey to brown scales; bulbose to turpinate, 1–2.5 cm thick, covered with darker grey to nearly black scales at upper portion. Annulus superior. Basidiospores 7–9 × 5.5–7 µm, broadly elliptical to elliptical, smooth, hyaline, amyloid.

In summer to autumn, scattered or **gregarious** on ground of coniferous and broad-leaved forests. Slightly poisonous, not edible. Distributed widely in China.

944 *Amanita fuliginea* Hongo

Pileus 3–5 cm broad, dark grey to nearly black, with darker and looming fibrillose patterns or macules. Context white. Stipe 6–10 cm long, 0.5–1 cm thick, white to light grey, often covered with light brown squamules; bulbose at base, 1–2.5 cm thick. Volva shallowly cup-shaped, white to dirty white. Annulus grey, membranous. Basidiospores

7–9 × 6.5–8.5 µm, globose to subglobose, smooth, hyaline, amyloid.

In summer to autumn, on ground of Fagaceae and Pinaceae forests. Deadly poisonous. Distributed in Central and Southern China. Context white to greyish white.

945 *Amanita griseofolia* Zhu L. Yang

Pileus 3–7 cm broad, grey to brownish grey, with grey to dark grey pulverulent granules or felty scales, ribbed at margin. Context white to greyish white. Lamellae light grey when mature, becoming grey to dark grey when dried. Stipe 8–16 cm long, 0.5–1.5 cm thick, white to dirty white, covered with grey fibrillose scales. Annulus abscent. Volva grey to dark grey, pulverulent. Basidiospores 10–13.5 × 9.5–13 µm, globose to subglobose, smooth, hyaline, inamyloid.

On ground of Fagaceae and Pinaceae mixed forests. Edible. Distributed widely in China.

946 *Amanita griseoverrucosa* Zhu L. Yang

Pileus 7–13 cm broad, light grey to dirty white, covered with light grey to grey, warty to pyramidal scales. Context thick, white, fleshy. Stipe 6–12 cm long, 0.7–2.5 cm thick, dirty white to light grey; ventricose to fusiform at base, 1.5–4 cm thick. Annulus fragile and fagacious. Basidiospores 8–11 × 5.5–7 µm, broadly elliptical to elliptical, smooth, hyaline, amyloid.

In summer to autumn, on ground of coniferous, broad-leaved and mixed forests. Toxicity unknown. Distributed in Central and Southern China.

947 *Amanita hemibapha* (Berk. & Broome) Sacc. s.l.

Pileus 6–15 cm broad, red to orange at disc, yellow at margin. Context thick at stipe, white to tincted with pileus colour. Lamellae light yellow to yellow. Stipe 8–15 cm long, 1–3 cm thick, pale yellow to yellow, covered with yellow to orange red irregular slough-shaped scales. Volva saccate, 2–5 cm high, 2–5 cm thick, white. Annulus light yellow to orange, superior. Basidiospores 8–9.5 × 6.5–8 µm, subglobose to broadly elliptical, smooth, hyaline, inamyloid.

In summer to autumn, on ground of coniferous, broad-leaved and mixed forests. Edible. Distributed widely in China.

948 *Amanita hongoi* Bas

Pileus 9–16 cm broad, hemispherical, oblate hemispherical to plane, slightly prominent at centre, light yellow to dirty yellow, becoming white to dirty white at outwards, with greyish white to light brown pyramidal or conical scales. Context thick, white. Lamellae free to subfree, close, white with pinkish yellow tinge, pulverulent at edge. Stipe 11–13 cm long, 1–3 cm thick, cylindrical, becoming wider downwards, white to dirty white, covered with white to light dark squamules; base enlarged, 2–7 cm thick, ventricose to bulbous. Annulus superior to nearly so, white to beige; radially rugulose on upper surface; warty on lower surface, membranous, fagacious, occasionally persistent. Basidiospores 7.5–9.5 × 6.5–8.5 µm, broadly elliptical to subglobose, smooth, hyaline, amyloid.

In summer to autumn, on ground of broad-leaved and mixed forests. Distributed in Central and Southern China.

949 ***Amanita ibotengutake*** T. Oda et al.

Pileus 6–15 cm broad, globose to oblate hemispherical when young, convex to plane at maturity, glabrous, slimy when wet, covered with white to greyish white blocky scales; scales partially falling off after raining or with age; margin striate, brown to brownish grey. Lamellae white to dirty white, close, free, unequal. Stipe 7–12 cm long, 1–2 cm thick, cylindrical, bulbosely enlarged at base, covered with fibrillose scales; stout and solid when young; stipe context becoming soft and hollow at maturity, greyish white. Volva on enlarged base of stipe. Basidiospores 10–12 × 7–9.5 µm, broadly elliptical, smooth, hyaline, inamyloid.

In summer to autumn, solitary or scattered on ground of broad-leaved and mixed forests. Poisonous. Distributed in Northeast China.

950 ***Amanita kotohiraensis*** Nagas. & Mitani.

Pileus 5–8 cm broad, subhemispherical when young, convex to plane with age, white, sometimes with beige tinge at disc, often with blocky velar remnants; often appendulate with flocculence at margin. Context white, unchanging when injured, with a pungent odour. Lamellae free, light yellow, close. Stipe 6–13 cm long, 0.5–1.5 cm thick, subcylindrical, white, covered with white squamules; enlarged at base, 1.5–4 cm thick, subglobose, zonately umbonate, often underground. Annulus superior to subapical, white, membranous, persistently suspending at margin, or fagacious. Basidiospores 7.5–9.5 × 5–6.5 µm, broadly elliptical, smooth, hyaline, amyloid.

In summer to autumn, on ground of broad-leaved and mixed forests. Poisonous. Distributed in Central and Southern China.

951 *Amanita longistriata* S. Imai

Pileus 3–9 cm broad, hemispherical when young, plane with age, slightly prominent at centre, grey, light grey or light yellow at margin, sometimes with pinkish tinct, greyish brown to brown at disc, radially striate. Context white. Lamellae 8–11 mm broad, pink at first, light pink with age, sinuate, close, unequal. Stipe 9–18 cm long, 0.5–1.5 cm thick, subcylindrical, tapering at apex, hollow, pink at first, becoming white with age. Annulus superior, membranous. Volva white, saccate. Basidiospores 8–13 × 7.5–11 μm, broadly elliptical, smooth, hyaline, inamyloid.

In summer, solitary on ground of broad-leaved and mixed forests. Distributed in Northeast and Northern China.

952 *Amanita macrocarpa* W.Q. Deng et al.

Pileus 15–40 cm broad, subglobose when young, then subhemispherical, oblate hemispherical, convex to plane, upturned at margin at maturity, slightly with orange to orange red tinge when young, dirty white to brownish orange or light brown with age, covered with numerous pale dirty brown to brownish orange (sometimes pinkish) pyramidal scales or warts 2–3 mm in height. Context thick, white, pale yellow when injured. Lamellae free, white to beige, crowded. Stipe 18–35 cm long, 2–5 cm thick, enlarged to 5.7–6.2 cm thick at base, cylindrical to obclavate, dirty white to light orange yellow, with squamules from white to reddish to yellowish brown at base; stipe context white, becoming yellowish when injured. Annulus medial to superior, dirty white to beige, thick membranous, fagacious. Basidiospores 7–9 × 5–6 μm, elliptical, smooth, hyaline, amyloid.

In spring to summer, scattered on ground in broad-leaved forests. Edibility unknown. Distributed in Southern China.

953 *Amanita manginiana* Har. & Pat.

Pileus 5–15 cm broad, grey, dark grey to brown, with darker fibrillose looming patterns or macules, often appendiculate with white velar remnants at margin. Context white. Lamellae free, close, unequal, white. Stipe 8–15 cm long, 0.5–3 cm thick, white, covered with white fibrillose to pulverulent scales, ventricose to clavate at base. Volva shallowly saccate, white to dirty white. Annulus apical to subapical, membranous, fragile, white, fagacious. Basidiospores 6–8 × 5–7 μm, subglobose to broadly elliptical, smooth, hyaline, amyloid.

In spring to summer, scattered on ground in coniferous or broad-leaved forests. Edible. However, it should be careful to distinguish it from the similar poisonous species. Distributed in Central China.

954 *Amanita muscaria* (L.) Lam.

Pileus 5.7–7.4 cm broad, hemispherical at first, plane to flat with age, sometimes depressed at centre, slimy when fresh, faintly or vaguely short ribbed at margin; darker and orange to red at disc, light orange to beige yellow at margin. Scales on pileus pyramidal or verrucose, white to light yellow. Context white. Lamellae free, close, unequal, white, beige when dried; lamellulae truncated near stipe. Stipe 12–13.5 cm long, 0.4–1 cm thick, cylindrical, tapering from base to apex; enlarged at base, bulbous to subbulbous, 1.5–2 cm thick; often with irregular scales at lower portion; stipe context soft to hollowing, white, brown when dried, with white scales. Annulus superior, membranous, white, persistent. Basidiospores 9–11 × 6–9 µm, broadly elliptical to elliptical, rarely subglobose, smooth, hyaline, inamyloid.

In spring to summer, gregarious or scattered on ground in coniferous, mixed or broad-leaved forests (Fagaceae and Pinaceae). Poisonous. Distributed in Inner Mongolia, Northeast and Northwest China.

955 *Amanita neoovoidea* Hongo

Pileus 5–13 cm broad, hemispherical, oblate hemispherical to plane, white to dirty white, slightly slimy when wet, estriate, pulverulent, usually with a yellowish white to light yellow universal veil, often with lacerated velar remnants at margin. Context white, slightly thick, changing to darker and reddish when injured. Lamellae free, close, unequal, white, white with earthy brown tinge, pruinose-granular at edge. Stipe 8–14 cm long, 1.2–2.2 cm thick, cylindrical, subfusiform at base, white to dirty white, with pulverulent scales; stipe context solid or soft, nearly white. Annulus superior, membranous, fragile and fagacious. Volva yellowish white to light yellow (concolourous with velar remnants on pileus). Basidiospores 8–10.5 × 6–8.5 µm, elliptical to elliptical-oblong, smooth, hyaline, amyloid.

In spring to summer, on ground in forests. Poisonous. Distributed in Central China.

956 *Amanita oberwinklerana* Zhu L. Yang & Yoshim. Doi

Pileus 3–6 cm broad, white, sometimes beige, glabrous or sometimes with 1–3 large patches of white membranous velar remnants. Context white, unchanging when injured. Lamellae free, close, white. Stipe 5–7 cm long, 0.5–1 cm thick, cylindrical, bulbous at base, 1–2 cm thick. Annulus superior, membranous. Volva shallowly cup-shaped to saccate, or nearly abscent. Basidiospores 8–10.5 × 6–8 µm, elliptical, smooth, hyaline, amyloid.

In spring to summer, on ground in mixed forests. Poisonous. Distributed in Central and Southern China.

957 *Amanita orientifulva* Zhu L. Yang et al.

Pileus 5–15 cm broad, plane, reddish brown to brown or dark brown. Context white, unchanging when injured. Lamellae free, subdistand to subclose, unequal, white. Stipe 8–15 cm long, 0.5–3 cm thick, dirty white to light brown, covered with reddish brown to greyish brown scales. Annulus abscent. Volva saccate, 4–6 cm high, 1.5–5 cm thick, white with rusty spots on surface. Basidiospores 10–14 × 9.5–13 µm, globose to subglobose, smooth, hyaline, inamyloid.

In spring to summer, on ground in coniferous, mixed or broad-leaved forests. Distributed widely in China.

958 *Amanita orsonii* Ash. Kumar & T.N. Lakh

Pileus 3–12 cm broad, reddish brown, yellowish brown to greyish brown; velar remnants on surface cracked into subpyramidal, verrucose, granular to flocculent scales, dirty white, light grey to greyish brown. Context white, becoming light reddish brown when injured. Lamellae free, subclose, unequal, white. Stipe 7–13 cm long, 0.5–1.5 cm thick, bulbous at base. Volva as concentric zones at upper half of bulbous stipe base. Annulus superior, membranous, concolourous with stipe. Every part of basidioma becoming reddish brown when injured. Basidiospores 7–9 × 5.5–7.5 µm, broadly elliptical to elliptical, smooth, hyaline, amyloid.

In spring to summer, solitary or gregarious on ground in coniferous or mixed forests. Poisonous when eaten raw. Distributed widely in China.

959 *Amanita ovalispora* Boedijn

Pileus 4–7 cm broad, grey to dark grey, glabrous or occasionally with white velar remnants, long ribbed at margin. Context white, unchanging when injured. Lamellae free, unequal, white, becoming grey or light brown when dried. Stipe 6–10 cm long, 0.5–1.5 cm thick, often with white pulverulent scales on upper half. Annulus abscent. Volva saccate to shallowly, membranous. Basidiospores 9–11 × 7–9 µm, broadly elliptical to elliptical, smooth, hyaline, inamyloid.

In summer to autumn, scattered on ground in broadleaved forests. Distributed in Central and Southern China.

960 *Amanita pallidorosea* P. Zhang & Zhu L. Yang

Pileus 4–11 cm broad, broadly campanulate at first, becoming plane slightly prominent at centre, white at margin, pale rose at disc at first, becoming pinkish white and finely radiately striate in later stage. Context white, unchanging when injured. Lamellae sinuate, close, unequal, white. Stipe 8–15 cm long, 0.6–1.2 cm thick, subcylindrical, tapering upwards, white, with minute fibrillous squamules, enlarged at base. Annulus superior, white, membranous. Volva saccate, white, persistent. Basidiospores 6–10 × 6–9 µm, globose to subglobose, smooth, hyaline, amyloid.

In summer, solitary to gregarious on ground in broadleaved forests. Deadly poisonous. Distributed in Northeast and Northern China.

961 *Amanita parvipantherina* Zhu L. Yang et al.

Pileus 3–6 cm broad, light yellow to yellow, covered with dirty white, light yellow to pale grey verrucose to pyramidal velar remnants. Context white, unchanging when injured.

Lamellae free, unequal, white. Stipe 4–10 cm long, 0.5–1 cm thick, light yellow, beige to white; enlarged at base, bulbous to oval, 1–2 cm in diameter. Volva white, light yellow or pale grey, at stipe base. Annulus small, superior. Basidiospores 8.5–11.5 × 7–8.5 μm, broadly elliptical to elliptical, smooth, hyaline, inamyloid.

In summer to autumn, on ground in mixed forests. Suspected poisonous. Distributed widely in China.

962 *Amanita porphyria* Alb. & Schwein

Pileus 4–9 cm broad, subhemispherical to oblate hemispherical at first, then plane, slightly prominent at centre, estriate at margin, greyish brown at first, then becoming dark brown, with purplish brown scales, slimy when wet. Context white, thin. Lamellae free, close, unequal, white. Stipe 6–12 cm long, 1–1.5 cm thick, cylindrical or abruptly bulbous. Volva shallowly saccate, pale grey where connected with stipe, with shaggy scales. Basidiospores 8–9 × 7–9 μm, subglobose, smooth, hyaline.

In summer to autumn, solitary to scattered on ground in forests. Distributed in Northeast China.

963 *Amanita pseudogemmata* Hongo

Pileus 4–9 cm broad, flat to plane, covered with verrucose to pulverulent or sometimes felty velar remnants, variable in colour, from yellow, sulfur yellow to light brown in paler specimens, from orange brown, olive brown to grey brown in darker specimens (especially when the powder fallen off), lighter and ribbed at margin. Context white to light yellow. Lamellae free, beige, close. Stipe 6–10 cm long, 0.5–1.5 cm thick, subcylindrical or tapering upward, concolourous with pileus, when young nearly white to beige at upper portion, covered with yellow to brown scales, white and soft insides; base 1.5–4 cm thick, abruptly bulbous. Volva abruptly bulbous to shallowly saccate, often with a collar at upper portion, white to light yellow. Annulus superior, membranous, yellow on upper surface, light yellow on lower surface, membranous. Basidiospores 6.5–9.5 × 6–8.5 µm, broadly elliptical to subglobose, smooth, hyaline, inamyloid.

In summer to autumn, on ground in broad-leaved forests. Distributed in Central and Southern China.

964 *Amanita pseudoporphyria* Hongo

Pileus 4–12 cm broad, hemispherical when young, then becoming flat, plane to upturned at margin, brownish grey, darker at centre, glabrous, with looming fibrils and fibrillose patterns, subslimy, sometimes with velar remnants; glabrous and estriate at margin, often with white flocculent remnants of annulus. Context white, unchanging when injured, slightly thicker at centre. Lamellae free, white, close, unequal. Stipe 5–12 cm long, 0.6–1.8 cm thick, white, often with fibrillous scales or with white flocculence, enlarged downwards and then radicant at base, solid. Annulus superior, white, membranous. Volva bracted or saccate, white. Basidiospores 7.5–9 × 4–6 µm, oval to broadly elliptical, smooth, hyaline, amyloid.

In summer to autumn, on ground in coniferous or broad-leaved forests. Some people eat it, but it is reported that it contains a little of toxin, so it is not suitable for eating. Distributed in Central China.

965 *Amanita pseudovaginata* Hongo

Pileus 3–6 cm broad, oblate hemispherical, convex to plane, often depressed at disc and umbonate at centre; grey, light grey to greyish brown, lighter at margin, sometimes whole pileus nearly white; glabrous, or with light grey to dirty white velar remnants, long ribbed or sulcate at margin. Context white. Lamellae free, white, sometimes light grey when dried, unequal. Stipe 5–12 cm long, 0.5–1.5 cm thick, subcylindrical, tapering upward, nearly white to greyish white, subglabrous, hollow, not enlarged at base. Volva 1.5–2 cm high, 1–2.5 cm thick, saccate or shallowly so, membranous, fragile, white to greyish white on outer surface, sometimes light grey; often with yellow brown macules; light grey to whitish on inner surface. Annulus abscent. Basidiospores 9.5–12.5 × 8–10.5 µm, subglobose to broadly elliptical, smooth, hyaline, inamyloid.

In summer to autumn, on ground in mixed forests composed of *Pinus yunnanensis*, *Pinus massoniana* or *Pinus massoniana* and *Quercus* spp.. Distributed in Northern, Northwest, Central and Southern China.

966 *Amanita rubescens* Pers.

Pileus 2–5.9 cm broad, oblate hemispherical to plane, prominent at centre, unconspicuously striate at margin, sticky when fresh; bright yellow, pinkish, greyish or dull brown, darker at disc, covered with white felty scales; scales fagacious. Context white, becoming brown when injured. Lamellae 3–5 mm broad, free, white to whitish, becoming brown, yellowish brown when dried, subclose. Stipe 5–8 cm long, 1–3 cm thick, subcylindrical, hollow, with fibrillose scales, concolourous with pileus, enlarged to bulbous at base. Annulus white, membranous, fagacious. Volva composed of greyish brown flocculent scales. Basidiospores 7–8.5 × 5.5–7 μm, broadly elliptical, elliptical to oval, smooth, hyaline, amyloid.

In summer to autumn, solitary or scattered on ground in coniferous and broad-leaved forests. Edible, however, it has also been reported that it contains some hemolytic substances, so it cannot be eaten raw. Distributed in Central and Southern China.

967 *Amanita rubrovolvata* S. Imai

Pileus 2–6.5 cm broad, red to orange red, orange to yellowish at margin, covered with red, orange red to yellow pulverulent to granular velar remnants. Context white, unchanging when injured. Lamellae free, subclose, white. Stipe 5–10 cm long, 0.5–1 cm thick, enlarged to bulbous at base, covered with red to orange red pulverulent velar remnants on upper half. Annulus superior. Basidiospores 7.5–9 × 7–8.5 μm, globose to subglobose, smooth, hyaline, inamyloid.

In summer to autumn, on ground in forests. Poisonous. Distributed widely in China.

968 *Amanita rufoferruginea* Hongo

Pileus 4–7 cm broad, yellowish brown, covered with lateritious red, orange red to buff brown velar remnants. Context white, unchanging when injured. Lamellae white. Stipe 7–10 cm long, 0.5–1 cm thick, densely covered with lateritious red, rubigenous red pulverulent scales, enlarged at base, 1.5–2 cm thick, covered with flocculent to pulverulent velar remnants at upper portion. Annulus superior, fragile. Basidiospores 7–9 × 6.5–8.5 μm, subglobose, smooth, hyaline, inamyloid.

In summer to autumn, scattered on ground in broad-leaved forests. Poisonous. Distributed in Central and Southern China.

969 *Amanita sculpta* Corner & Bas

Pileus 8–18 cm broad, subhemispherical when young, incurved at margin, greyish brown, light brown to purplish brown; velar remnants pyramidal to verrucose, greyish brown to dark brown, more fagacious at margin; margin estriate, often with flocculent appendages. Context white to light brown, becoming brown to dark brown when injured. Lamellae free to subfree, white tinged with reddish at first, becoming purplish brown, dark purplish brown to black when dried, unequal. Stipe 8–20 cm long, 1–3 cm thick at upper portion, cylindrical, often tapering toward apex, 2.5–6 cm thick at base, subfusiform. Annulus superior, fragile. Basidiospores 8–11 × 8–10.5 μm, globose to subglobose, smooth, hyaline, inamyloid.

In summer to autumn, on ground in forests. Suspected poisonous. Distributed in Central and Southern China.

970 *Amanita* cf. *similis* Boedijn

Pileus 8–20 cm broad, oblate hemispherical to plane, umbonate at centre, glabrous, slimy when wet, ochraceous to yellow, brown at central umbo, ribbed at margin. Context white, light yellow at pileal surface. Lamellae free to subfree, white to seldom light yellow, close, unequal; lamellulae often truncated at inward end. Stipe 9–18 cm long, 1–3.5 cm thick, subcylindrical or slightly tapering toward apex, light yellow to yellow, covered with yellowish brown to yellow snake-sloughy scales, hollow, not enlarged at base. Annulus light yellow to light orange, membranous, superior. Volva 2–5 cm high, 2–5 cm broad, saccate, white. Basidiospores 9–12.5 × 7–9 µm, broadly elliptical to elliptical, smooth, hyaline, inamyloid.

In summer to autumn, on ground in coniferous forests or mixed coniferous and broad-leaved forests. Distributed in Central China.

971 *Amanita sinensis* Zhu L. Yang

Pileus 7–12 cm broad, campanulate or hemispherical at first, becoming oblate hemispherical to plane with age, obviously ribbed or not at margin, greyish white to grey, covered with grey, dark grey to greyish brown universal veil, with verrucose to granular scales at centre, verrucose to flocculent at margin; scales usually partially fagacious. Context thin, white. Lamellae free to subfree, close, unequal, white. Stipe 8–15 cm long above ground, 1–2.5 cm thick, subcylindrical, dirty white to light grey, with light grey, grey to dark grey pulverulent to flocculent and fagacious scales, clavate to subfusiform at base, with a long subterranean radicant base. Annulus apical to subapical, membranous, fagacious. Basidiospores 9.5–12.5 × 7–8.5 µm, broadly elliptical to elliptical, smooth, hyaline, amyloid.

In summer to autumn, on ground in coniferous or broad-leaved forests. Distributed in Central and Southern China.

972 *Amanita spissacea* S. Imai

Pileus 3.5–12 cm broad, hemispherical at first, becoming plane at age, slimy when wet, grey to greyish brown, inconspicuously striate at margin, with blackish brown pyramidal or granular scales; scales dense, fagacious. Context white. Lamellae free, close, unequal, white. Stipe 3–12 cm long, 1.5–2 cm thick, cylindrical, darker above annulus, grey below annulus, with grey fibrillose scales, enlarged at base. Annulus superior, membranous, white on upper surface, grey on lower surface, blackish grey at margin. Volva as granular zones. Basidiospores 7.5–9 × 5.6–7.5 µm, broadly elliptical, smooth, hyaline, dextrinoid.

In spring to autumn, solitary or gregarious on ground in coniferous or broad-leaved forests. Poisonous. Distributed in Northern, Central and Southern China.

973 *Amanita subglobosa* Zhu L. Yang

Pileus 4–10 cm broad, light brown to amber brown; velar remnants white to light yellow, forming pyramidal to verrucose scales. Stipe 5–15 cm long, 0.5–2 cm thick, cylindrical; base bulbous, 1.5–3.5 cm thick, with a small granular to pulverulent volva at upper portion, collar-like. Annulus superior, membranous. Basidiospores 8.5–12 × 7–9.5 µm, broadly elliptical to elliptical, smooth, hyaline, inamyloid.

In summer to autumn, on ground in mixed forests composed of *Pinus*, *Populus* and Fagaceae. Suspected poisonous. Distributed widely in China.

974 *Amanita subjunquillea* S. Imai

Pileus 3–6 cm broad, yellowish brown, dirty orange to mustard yellow. Context white, yellow at pileus surface, unchanging when injured. Lamellae free, unequal, white. Stipe 4–12 cm long, 0.3–1 cm thick, cylindrical, white to light yellow; bulbous at base, 1–2 cm thick. Annulus subapical to superior, white. Volva shallowly saccate, white to dirty white. Basidiospores 6.5–9.5 × 6–8 µm, globose to subglobose, smooth, hyaline, amyloid.

In summer to autumn, on ground in forests. Deadly poisonous. Distributed widely in China.

975 *Amanita sychnopyramis* f. *subannulata* Hongo

Pileus 3–8 cm broad, plane, light brown to dark brown, scaly; scales pyramidal to conical, white to light grey, darker at base. Context white, unchanging when injured. Lamellae free, unequal, white. Stipe 5–11 cm long, 0.7–1.5 cm thick, cylindrical; base enlarged, bulbous to ventricose at base. Volva at upper portion of stipe base, verrucose, granular to pulverulent. Annulus medial to inferior. Basidiospores 6.5–8.5 × 6–8 μm, globose to subglobose, smooth, hyaline, inamyloid.

In summer to autumn, on ground in broad-leaved or mixed forests. Poisonous. Distributed in Central and Southern China.

976 *Amanita vaginata* (Bull.) Lam.

Pileus 3–8 cm broad, convex to plane, grey, sometimes light brown, obviously striate at margin, sometimes with large patchy remnants of universal veil. Context white, unchanging when injured. Lamellae free, nearly white. Stipe 5–10 cm long, 0.5–1.5 cm thick, white to dirty white, subglabrous to with light grey to light brown fibrillose scales. Annulus abscent. Volva saccate to shallowly so. outer surface white to dirty white, inner surface white. Basidiospores 9.5–11 × 9–10.5 μm, globose to subglobose, smooth, hyaline, inamyloid.

In summer to autumn, on ground in forests of *Pinus* and Fagaceae. Inedible when raw. Distributed widely in China.

977 *Amanita vestita* Corner & Bas

Pileus 3–7 cm broad, convex to plane, estriate, covered with yellowish brown, light brown to dark brown shaggy, flocculent to felty veil, with verrucose fagacious scales at centre, often flocculent at margin. Context white, unchanging when injured. Lamellae free to subfree, white. Stipe 4–6 cm long, 0.5–1 cm thick, subcylindrical, thicker to base, greyish white or grey, covered with nearly white to light grey fibrillose to flocculent scales. Stipe solid, white; enlarged at base, 1–2 cm thick, subfusiform, with a short radicant base. Annulus superior, fragile and fagacious. Basidiospores 7–10.5 × 5–7 µm, elliptical, smooth, hyaline, amyloid.

In summer to autumn, scattered on ground in forests of tropical and southern subtropical zones. Distributed in Southern China.

978 *Amanita virgineoides* Bas

Pileus 7–15 cm broad, hemispherical to plane, white, with conical to pyramidal scales. Stipe 10–20 cm long, 1.5–3 cm thick, cylindrical, white, covered with white flocculent to pulverulent scales; ventricose to oval at base. Volva indistinct, as verrucose to granular prominences. Annulus fragile, with minute verrucose to conical prominences on lower surface. Basidiospores 8–10 × 6–7.5 µm, broadly elliptical to elliptical, smooth, hyaline, amyloid.

In summer to autumn, on ground in mixed forests. Poisonous. Distributed widely in China.

979 *Amanita virosa* (Fr.) Bertill.

Pileus 2.4–5.3 cm broad, broadly campanulate to oblate hemispherical at first, becoming plane, sometimes slightly prominent at centre, pure white, slightly yellowish at prominent disc, slimy when fresh; margin inconspicuously striate, incurved, white, sometimes with greyish tinge. Context white. Lamellae free, close, white. Stipe 7–15 cm long, 0.4–1.2 cm thick, cylindrical, soft insides, enlarged to bulbous at base, white, with white squamules. Annulus superior or apical, membranous, white. Basidiospores 8.5–12.5 × 7.5–11.5 μm, globose to subglobose, rarely broadly elliptical, smooth, hyaline, amyloid.

In summer to autumn, solitary or scattered on ground in coniferous and mixed forests. Serious poisonous. Distributed in Northern and Northeast China.

This species is similar to *A. subjunquillea* var. *alba*. The latter species has smaller, globose to subglobose basidiospores, and its volva is composed of filamentous and inflated cells. Whereas, the basidiospores of this species are subglobose to broadly elliptical, and there is no inflated cells in its volva.

980 *Amanita yuaniana* Zhu L. Yang

Pileus 7–13 cm broad, subplane or slightly plano-umbonate, grey to snuff brown, with white to dirty white looming patterns, and with looming radial fibrils, short sulcate at margin. Context white, unchanging when injured. Lamellae free, white. Stipe 7–14 cm long, 1–2.5 cm thick, cylindrical, white to light grey. Volva saccate to shallow so. Annulus superior, fragile. Basidiospores 9.5–12 × 6.5–8 μm, elliptical, smooth, hyaline, inamyloid.

In summer to autumn, on ground in coniferous or mixed forests. Edible. Distributed in Central China.

981 *Amanita zangii* Zhu L. Yang et al.

Pileus 5–6 cm broad, oblate hemispherical at first, becoming flat to plane at maturity, white to beige, with greyish brown, dark grey to nearly black velar remnants; velar remnants often cracked into zonately arranged felty, subfibrillose, subconical or verrucose scales; margin not sulcate, often flocculent. Context white. Lamellae free, close, unequal, white. Stipe 6–8 cm long, 0.8–1.2 cm thick, subcylindrical, white to beige; base slightly enlarged, 1.5–2 cm thick, subclavate to bulbous, with white to grey pulverulent to flocculent velar remnants. Annulus apical to subapical, white, fragile. Basidiospores 8.5–11.5 × 6.5–8 μm, elliptical, amyloid.

In summer to autumn, on ground in forests of tropical zone. Distributed in Southern China.

982 *Ampulloclitocybe clavipes* (Pers.) Redhead et al.

Pileus 3.5–7 cm broad, plane or slightly depressed when young, becoming centrally depressed or infundibuliform, usually with a small umbo at centre, greyish brown or dark brown when fresh, darker at centre, glabrous or tomentose, often incurved at margin when young. Context white, thick. Lamellae decurrent, thin, white, creamy or light yellow, subdistant or close, unequal. Stipe 3–6.5 cm long, 0.6–1.4 cm thick, cylindrical or subclavate, fibrillosely striate, concolourous with pileus or lighter, solid. Basidiospores 6–9.5 × 4–5.5 μm, subglobose to elliptical, smooth, hyaline.

In summer to autumn, solitary or scattered on ground or debris in forests. Edible, but it is also recorded as slightly poisonous. Eaters are more likely to be poisoned when they drink alcohol at the same time. Therefore, attention should be paid when taking this mushroom. Distributed in Northeast China, Qinghai-Tibet Region, and Inner Mongolia.

983 *Anthracophyllum nigritum* (Lév.) Kalchbr.

Pileus 0.6–3.5 cm long, 0.5–3 cm broad, subreniform, semicircular, flabellate or elliptical or conchate, radially sulcate, fleshy brown, tawny or reddish tawny, sometimes paler at margin. Context thin, fairly tough, brown to chartreuse. Lamellae distant, narrow, with 9–13 complete lamellae and some lamellulae, radiating from a base of attachment, concolourous with pileus, or grey, greyish brown to dark brown, seldom furcate. Stipe abscent or very rudimentary, lateral. Basidiospores 6.5–9 × 4–5 μm, elliptical, hyaline to light brown.

In summer to autumn, gregarious on dead twigs or branches of broad-leaved trees. Distributed in Southern China.

984 *Armillaria mellea* (Vahl.) P. Kumm.

Pileus 3–7 cm broad, oblate hemispherical to plane, yellow to yellowish brown, covered with brown to brown scales, dense at centre. Context nearly white to light yellow, unchanging when injured. Lamellae adnate to subdecurrent, nearly white to light yellow or brownish, lighter than pileus. Stipe 5–10 cm long, 0.3–1 cm thick, cylindrical, white above annulus, greyish brown below annulus, with greyish brown scales. Annulus superior, white on upper surface, light brown on lower surface. Basidiospores 8.5–10 × 5–6 μm, elliptical to elliptical-oblong, smooth, hyaline, inamyloid.

In summer to autumn, on trees or rotten woods. Edible, cultivable. Distributed widely in China.

985 *Armillaria ostoyae* (Romagn.) Herink

Pileus 3–12 cm broad, convex at first, reddish brown to deep brown, then becoming plane and paler, yellowish brown at margin, light brown at disc, with light brown fibrillose scales, with less scales at margin. Context white to dirty white, sometimes becoming light brown when mature. Lamellae adnate to decurrent, white or dirty white at first, becoming light brown when mature, with fibrillose scales, with yellow mycelium at base. Annulus present. Basidiospores 8–11 × 5–7 μm, elliptical, smooth, hyaline, inamyloid.

In summer to autumn, gregarious on ground of coniferous forests or base of trunks. Edible. Distributed widely in China.

986 *Armillaria tabescens* (Scop.) Emel

Pileus 2.8–8.5 cm broad, oblate hemispherical when young, becoming plane, upturned at margin when mature, melline yellow or yellowish brown, rust brown at maturity, usually darker and with fibrillose squamules, not slimy. Context white or slightly cream. Lamellae white to dirty white, or with darker fleshy pink tinge, decurrent, subdistant, unequal. Stipe 2–13 cm long, 3–9 mm thick, cylindrical, dirty white on upper portion, greyish brown to blackish brown on lower portion, sometimes twisted, with appressed silky fibrils, soft to hollow insides. Annulus abscent. Basidiospores 7.5–10 × 5–7.5 μm, broadly elliptical to oval, smooth, hyaline, inamyloid.

In summer to autumn, fasciculate on deadwood, trunk bases and rhizosphere of broad-leaved trees in forests. Edible. Distributed widely in China.

987 *Arrhenia epichysium* (Pers.) Redhead et al.

Pileus 1.5–3.5 cm broad, convex-depressed to infundibuliform, incurved and then decurved at margin, translucently striate, hygrophanous, grey-brown to dingy-brown, glabrous, furfuraceous at centre. Context thin, light brown, odour and taste mild. Lamellae decurrent, light grey. Stipe 1–3 cm long, 2–3 mm thick, cylindrical, subequal, usually white tomentose at base, glabrous elsewhere, light greyish brown. Basidiospores 7–8.5 × 4–4.5 μm, elliptical to elliptical-oblong, smooth, hyaline, inamyloid.

In summer to autumn, solitary or gregarious on deadwood in coniferous, broad-leaved or mixed forests. Edible. Distributed in Northern and Northeast China.

988 *Arrhenia spathulata* (Fr.) Redhead

Pileus 5–15 cm broad, conchate, spatulate to flabellate, radially tomentulose, especially at centre, deep greyish brown when wet, becoming lighter when dry. Context membranous. Lamellae veined, reticulate to lamellate, distant and even, greyish white. Stipe very short or rudimentary, tomentulose, greyish white. Basidiospores 7–9 × 4.5–5.5 µm, elliptical to subpyriform, smooth, hyaline, inamyloid.

Solitary or gregarious with moss on decayed wood. Distributed in Northeast China.

989 *Asterophora lycoperdoides* (Bull.) Ditmar

Pileus 5–30 mm broad, subglobose, becoming convex, white to light grey, brownish when mature, subglabrous to tomentulose at first, then producing yellowish white, dark yellow to dark brown powdery chlamydospores soon on the surface. Context thick, nearly white to brown, with unpleasant smell. Lamellae distant, white to greyish white, sometimes furcate, with numerous lamellulae. Stipe 1–4 cm long, 2–6 mm thick, cylindrical, white to light greyish brown, with white to brown mycelium at base. Basidiospores 5–6 × 3–4 µm, elliptical, smooth, hyaline. Chlamydospores 15–18 µm diam., with a subglobose main body, stellate in overall shape for having subconical and thick spines, yellow, numerous on pileus surface.

In summer to autumn, parasitic on fruiting bodies of macrofungi like *R. nigricans*-complex. Gregarious to fasciculate. Distributed widely in China.

990 *Baeospora myosura* (Fr.) Singer

Pileus 1–2.5 cm broad, glabrous, light yellowish brown to brown, becoming lighter when dried. Context thin, white to nearly white. Lamellae adnate, close, white. Stipe 3–5 cm long, 1–3 mm thick, cylindrical, slender, glabrous, lighter than pileus, with white mycelium at base. Basidiospores 4.5–6 × 2.5–3.5 μm, elliptical, smooth, hyaline, amyloid.

In late autumn to winter, on cones in forests. Distributed in Northeast China.

991 *Baeospora myriadophylla* (Peck) Singer

Pileus 0.8–3 cm broad, convex or plane, slightly concave at centre, becoming plane or lobed, glabrous, hygrophanous, greyish purple to dirty purple when young, greyish brown, purplish brown, to light grey at maturity. Context white, yellowish white, to slightly grey. Lamellae adnate, very close, light, greyish purple to dirty purple when young. Stipe 2–5 cm long, 1.5–4 mm thick, cylindrical, hollow, white pruinose at apex, light reddish grey when young, then glabrous and greyish purple at maturity, with tomentose mycelium at base. Basidiospores 3–4 × 2–2.5 μm, subglobose to elliptical, smooth, hyaline, amyloid.

In summer, on deadwood in coniferous forests. Distributed in Northeast China.

992 *Bolbitius titubans* (Bull.) Fr.

Pileus 1.5–5 cm broad, oval to subglobose at first, becoming broadly campanulate to broadly convex, then becoming plane, depressed at centre, brittle, slimy, yellow or light yellowish green, sometimes light brown to light grey, then fading, light yellow at disc, light grey or light brown at margin, glabrous, often plicated. Context light yellow, brittle. Lamellae free, close, brittle and soft, nearly white or light yellow, becoming cinnamon rust. Stipe 3–12 cm long, 2–4 mm thick, cylindrical, hollow, brittle, squamulose surface, pulverulent or subglabrous, white, sometimes light yellow. Basidiospores 10–16 × 6–9 μm, elliptical, truncated at apex, smooth, rust brown.

In summer to autumn, solitary, scattered or gregarious on dung or fertilized grassland. Distributed in Northeast, Northern, Central, Southern, Inner Mongolia and Northwest China.

Agarics 2: Species with Names Beginning with C-E

993 *Callistosporium luteoolivaceum* (Berk. & M.A. Curtis) Singer

Pileus 1.5–3 cm broad, plane or umbilicate, furfuraceous to glabrous, olive brown, olive yellow to dark yellow, yellowish brown to dark reddish brown, purplish red in KOH. Context thin, dirty white or dark white. Lamellae adnate, close, yellow or golden yellow, dark red to purplish red when dried. Stipe 2–5 cm long, 5–8 mm thick, cylindrical or clavate, cinnamon, yellow brown or concolourous with pileus, dark brown or reddish brown when old or dried, fibrillose, hollow, sometimes grooved. Odour not distinctive, or slightly fragrant; taste bitter. Basidiospores 3–3.5 × 5–6 μm, broadly elliptical, smooth, hyaline, inamyloid.

On deadwood in coniferous forests. Distributed in Northeast and Central China.

994 *Calocybe gambosa* (Fr.) Donk

Pileus 5–14 cm broad, subhemispherical at first, then convex to plane, glabrous, nearly white to light brown, incurved at margin. Context white, thick. Lamellae white, narrow, close. Stipe 3–8 cm long, 2–4 mm thick, white to light yellow, hollow. Basidiospores 5.5–6.5 × 3–4.5 μm, elliptical, smooth, hyaline.

In summer to autumn, gregarious on grassland. Edible. Distributed in Inner Mongolia, Northeast and Northern China.

© Central China Farmer's Publishing House 2025

Y. Li et al., *Atlas of Chinese Macrofungal Resources*, https://doi.org/10.1007/978-981-96-0502-6_3

995 *Calyptella capula* (Holmsk.) Quél.

Pileus 0.3–1.5 cm long, 0.2–1 cm wide, hanging cupulate, campanulate, pixy hat-shaped or invertedly infundibuliform, white to creamy white, becoming yellow at maturity, even at margin, membranous, smooth on inner and outer surface; margin wavy and outcurved. Stipe 2.5 mm long, 0.5 mm thick. Basidiospores 5–6 × 3–4 µm, elliptical-oblong, smooth, hyaline, inamyloid.

In summer to autumn, gregarious on rotten stalk of Gramineae plants. Distributed in Southern China.

996 *Campanella junghuhnii* (Mont.) Singer

Pileus 0.4–2 cm broad, semicircular to circular-flabellate, white to yellowish, becoming cream, light yellow to dark yellow when dried, glabrous; margin incurved at first, then plane. Context thin, tough. Lamellae yellowish white, radiating from an attachment base, strongly interveined between lamellae, reticulate, decurrent, entire at edge. Stipe 2–3 mm long, 1 mm thick, cylindrical or curved cylindrical, lateral or eccentric, fibrous. Basidiospores 8–10.5 × 5.5–7 µm, broadly elliptical to subventricose, smooth, hyaline, inamyloid.

In summer to autumn, gregarious on rotten bamboo in forests or rotten logs in broad-leaved forests. Distributed in Northeast, Southern and Central China.

997 *Campanella tristis* (G. Stev.) Segedin

Pileus 0.4–3 cm broad, semicircular to reniform, often bowl-shaped, white, cream or light grey, slightly tinged with blue green when young, becoming cream, light yellow, to dark yellow when dried, uneven, short tomentulose, incurved at margin. Context soft, thin, gelatinized, translucent. Lamellae distant, thin, decurrent, with 8–10 main lamellae radiating from attachment base or stipe, with lamellulae and strong interveins between lamellae arranged in a reticulate pattern, white to slightly with aeruginous ting. Stipe 2–3 cm long, 1 mm thick, cylindrical or curved cylindrical, lateral or eccentric, sometimes unobvious. Basidiospores 8–11 × 4.5–6 µm, broadly elliptical to subventricose, smooth, hyaline, inamyloid.

Clustered or gregarious on rotten wood or logs of broad-leaved trees in mixed forests. Distributed in Northeast China.

This species is confusable with *C. junghuhnii* when old or dried. However, the cystidia of the latter species are long clavate and often longer than 50 µm. Whereas the cystidia of this species are smaller, only 30 µm.

998 *Cantharellula umbonata* (J.F. Gmel.) Singer

Pileus 3–7 cm broad, subhemispherical to oblate hemispherical at first, concaved to umbilicate at centre, incurved at margin when young, becoming plane at maturity, appressedly squamulose, greyish brown, blackish brown or reddish brown when wet, light grey when dried. Context thin, greyish brown or white. Lamellae adnate to subdecurrent, waxy, subclose, unequal, cream to white. Stipe 4–11 cm long, 3–7 mm thick, cylindrical, soft to hollow, light greyish brown, white pruinose at upper portion, with white mycelium at base. Basidiospores 9.1–11.4 × 3.7–4.6 µm, elliptical-oblong to elliptical, smooth, hyaline, amyloid.

In summer to autumn, solitary or gregarious on moss in *Picea asperata*. Edible. Distributed in Northwest China.

999 *Catathelasma ventricosum* (Peck) Singer

Pileus 8–15 cm broad, subhemispherical to flat, dirty white, greyish white to light brown, dry or slimy when wet, sometimes cracked at maturity, incurved at margin. Context thick, white, unchanging when injured. Lamellae decurrent, white to beige. Stipe 8–15 cm long, 2.5–5 cm thick, cylindrical to fusiform, dirty white to light yellowish brown, tapering to base. Annulus thick, membranous. Basidiospores 9–12 × 4–6 µm, elliptical to cylindrical, smooth, hyaline, amyloid.

In summer to autumn, on ground in coniferous or mixed forests. Edible. Distributed widely in China.

1000 *Chaetocalathus liliputianus* (Mont.) Singer

Pileus 0.3–1 cm broad, laterally dorsal on substrate, inversely cupulate to flabelliform, membranous, pure white, dry, tomentose and more tomentose; sulcate at margin; margin expanded to slightly incurved. Context white, thin, odourless. Lamellae white, radiating from a central attachment point, close, unequal. Stipe abscent. Basidiospores 7.5–10 × 5–7 µm, broadly elliptical, smooth, hyaline, inamyloid.

Gregarious on rotten wood in broad-leaved forests. Distributed in Southern China.

1001 *Chlorophyllum agaricoides* (Czern.) Vellinga

Pileus 3–5 cm broad, oval to globose at first, dirty white to light yellow, umbonate at top. Stipe 1–2.5 cm long, 1–1.5 cm thick, enlarged toward base, Exoperidium single, 1–2 mm thick, glabrous and pinkish at first, then broken from stipe, scaly. Gleba insides yellowish green to brownish rust at first, becoming white or yellowish at maturity; chambers 1 mm broad, labyrinth, with lamellate plates. Basidiospores 6–8 × 6–7 μm, globose to subglobose, smooth, yellow.

In autumn, solitary or scattered on grassland. Medicinal. Distributed in Inner Mongolia, Northeast and Northwest China.

1002 *Chlorophyllum hortense* (Murrill) Vellinga

Pileus 3.5–5.5 cm broad, subcylindrical when young, becoming conical, then plane to plano-convex, with an obvious obtuse umbo at centre at maturity, covered with light yellow to yellowish brown shreds, darker at disc, lighter at margin, white between shreds, often with white tomentum at margin. Context white, unchanging when injured or becoming pinkish. Lamellae close, free, unequal, light greyish yellow, unchanging when injured. Stipe 3–5.5 cm long, 0.5–1.2 cm thick, enlarged at base, hollow, light white to brown, becoming darker to base, changing light red to reddish brown when injured. Annulus medial, membranous, cream to light ochraceous, fagacious. Basidiospores 8–10.5 × 5.6–7.2 μm, broadly elliptical to oval, thick wall, smooth, subhyaline to slightly yellow.

Scattered or gregarious on ground at edge of forests or roadside. Distributed in Southern China.

1003 *Chlorophyllum molybdites* (G. Mey) Massee

Pileus 5–25 cm broad, white, hemispherical, oblate hemispherical, plane at maturity, slightly umbonate at centre; pileipellis dark brown or light brown when young, gradually broken into scales; scales at centre larger and thicker, brownish purple, becoming less or fagacious at margin. Context white or pinkish, soft. Lamellae free, wide, unequal, dirty white at first, becoming greenish, brown or grey, pruinose at edge. Stipe 10–28 cm long, 1–2.5 cm thick, cylindrical, dirty white to light greyish brown, fibrous, glabrous above annulus, with white fibrils below annulus; slightly enlarged at base, hollow; stipe context becoming brown when injured, with a fragrant smell when dried. Annulus superior, membranous, movable. Basidiospores 8–12 × 6–8 μm, broadly oval to broadly elliptical, smooth, subhyaline to slightly yellow, slightly truncated at apex.

In summer to autumn, gregarious or scattered on lawn or on ground in *Musa nana* forests after raining. Poisonous. Distributed in Inner Mongolia and Southern China.

This species is one of most poisonous mushroom species in Southern China. It mainly causes serious gastrointestinal discomfort and can also cause damage to organs such as liver and nervous system. Cultivable for research.

1004 *Chlorophyllum neomastoideum* (Hongo) Vellinga

Pileus 6–10 cm broad, umbonate at centre; scales brown, larger at centre, smaller at margin. Context nearly white, becoming reddish brown when injured. Lamellae free, white to beige, becoming reddish brown when injured. Stipe 10–12 cm long, 0.6–1 cm thick, central. Light brown to dark brown, enlarged at base. Annulus superior, fagacious. Basidiospores 7–9 × 5–6 μm, subelliptical to amygdaliform, smooth, subhyaline, slightly truncated at apex.

In summer to autumn, gregarious on ground in bamboo groves or other forests. Poisonous, do not take it as food. Distributed in Central China.

1005 *Chromosera cyanophylla* (Fr.) Redhead et al.

Pileus 0.6–2.5 cm broad, broadly convex to plane, depressed at centre at maturity, glabrous, light greyish purple when young, becoming yellow to yellowish brown, white at margin. Context thin, yellowish white. Lamellae decurrent, distant to close, light lavender when young, light lavender or creamy white at maturity. Stipe 10–35 cm long, 1–2 mm thick, cylindrical, hollow, sometimes bulbous at base, light lavender at first, fading to yellow to yellowish brown at maturity, slightly tinged with lavender. Basidiospores 6–7.5 × 3–4 μm, elliptical, smooth, hyaline, inamyloid.

In summer to autumn, scattered on dead wood of coniferous trees, such as *Abies fabri* or *Pinus*. Distributed in Northeast China.

becoming orange red to red at first, and then to brown when injured. Lamellae white at first, then light yellowish green, becoming brownish green to greenish brown at maturity. Stipe 10–14 cm long, 1–2 mm thick, cylindrical, concolourous with pileus or whitish and paler. Annulus superior, membranous. Volva membranous, white. Basidiospores 6–7.5 × 4–5 µm, smooth, brown.

In summer to autumn, on roadside or ground in forests in tropical or south subtropical zones. Distributed in Southern China.

1006 *Cibaomyces glutinis* Zhu L. Yang et al.

Pileus 3–4.5 cm broad, oblate hemispherical, dirty white, grey to brown, prominent at centre. Context white. Lamellae white to beige. Stipe 3–9.5 cm long, 3–8 mm thick, cylindrical, white to light grey, slimy, covered with light brown felty scales. Basidiospores 10.5–14 × 9–11.5 µm, subglobose to broadly elliptical, with obvious digitate spines, hyaline.

On ground in broad-leaved forests, with a radicant base connected with dead wood of underground. Edible. Distributed in Central China.

1007 *Clarkeinda trachodes* (Berk.) Singer

Pileus 8–15 cm broad, oblate hemispherical, white to dirty white, with brown to chocolate scales, central scales often remaining complete and large, surrounding scales cracked into massive recurved squamules. Context nearly white,

1008 *Clitocybe costata* Kühner & Romagn.

Pileus 3–5 cm broad, plane when young, convex-depressed at centre to infundibuliform at maturity, striate and wavy at margin, tomentulose, thick tomentose or hirsute at centre, hygrophanous, ochraceous brown to beige brown, darker at disc. Context white, thin. Lamellae white to dirty creamy, wide, decurrent, furcate near stipe, entire to crenulated at edge. Stipe 3–4.5 cm long, 5–8 mm thick, cylindrical, smooth, ochraceous brown, with fine and white fibrils. Basidiospores 5–7 × 3.5–4.5 µm, elliptical, smooth, hyaline.

On leaf litter in mixed forests. Distributed in Northeast China.

1009 *Clitocybe dealbata* (Sowerby) P. Kumm.

Pileus 3–4 cm broad, hemispherical at first, then convex-depressed sometimes infundibuliform, white, light yellow, light yellowish brown, incurved or wavy at margin. Context white. Lamellae decurrent, close, white, yellowish white. Stipe 2–3.6 cm long, 0.5–1 cm thick, subcylindrical, white, slightly enlarged at base. Basidiospores 5–6 × 3.5–4 µm, subelliptical, smooth, hyaline, inamyloid.

In summer to autumn, gregarious or fasciculate on ground in forests. Poisonous. Distributed in Inner Mongolia and Northwest China.

This species is similar to *C. phyllophila*, whereas the basidiospores of the latter species are oval or subglobose.

1010 *Clitocybe gibba* (Pers.) P. Kumm.

Pileus 2–10 cm broad, oblate hemispherical at first, then plane, convex-depressed at centre to infundibuliform at maturity, often minutely umbonate or papillate at disc when young, dry, thin, light yellow to light brown, finely silky-pubescent at first, then glabrescent; margin sharp and wavy. Context thin, white. Lamellae decurrent, white, thin, close, narrow, unequal. Stipe 2–5 cm long, 0.5–1 cm thick, cylindrical, white, concolourous with pileus or lighter, glabrous, soft inside; base not enlarged to slightly enlarged, with white mycelium. Basidiospores 6–9 × 3.5–5 µm, suboval, elliptical, or long amygdaliform, smooth, hyaline, inamyloid.

In summer to autumn, solitary or gregarious on ground or leaf litter in broad-leaved or coniferous forests, or on grassland. Edible. Distributed in Inner Mongolia, Qinghai-Tibet Region, Northeast, Northern, and Northwest China.

1011 *Clitocybe nebularis* (Batsch) P. Kumm.

Pileus 4.5–12 cm broad, infundibuliform at first, becoming plane at maturity, light greyish brown to dirty white, darker at centre, even at margin. Context white, unchanging when injured. Lamellae decurrent, close, white at first, light yellow at maturity, unequal. Stipe 5–10 cm long, 1.5–4 cm thick, cylindrical, slightly enlarged at base, dirty white, solid to hollow. Basidiospores 5–7.2 × 3–4.5 µm, smooth, hyaline.

In summer to autumn, clustered to scattered on ground in forests. Edible, but also recorded as poisonous. Distributed in Inner Mongolia, Qinghai-Tibet region, Northeast, Northern, and Northwest China.

This species is similar to *C. odora*, but the pileus of the latter species is white with olivaceous tinge, and the context is thinner.

1012 *Clitocybe odora* (Bull.) P. Kumm.

Pileus 2–7 cm broad, hemispherical to oblate hemispherical at first, plane at maturity, depressed or prominent at centre, white, partially tinged with light yellowish green, often tinged with light yellow brown at disc; glabrous, estriate at margin. Context white. Lamellae adnate or subdecurrent, close, white, cream to pinkish white. Stipe 2–5 cm long, 1–1.5 cm thick, subcylindrical, white to yellowish white. Basidiospores 5.5–7 × 3.5–5 μm, oval to broadly elliptical, smooth, hyaline.

In summer to autumn, gregarious or scattered on ground in forests. Edible. Distributed in Northwest China.

1013 *Clitocybe phyllophila* (Pers.) P. Kumm.

Pileus 4.5–11 cm broad, oblate spherical at first, infundibuliform at maturity, white, with white tomentum, even at margin. Context white, unchanging when injured. Lamellae decurrent, close, white, unequal, entire at edge. Stipe 4–9 cm long, 0.4–1.2 cm thick, cylindrical, central, slightly twisted, tomentulose, hollow. Basidiospores 4.5–7 × 2.8–4 μm, elliptical, citriform, smooth, hyaline.

Gregarious on ground in broad-leaved forests. Poisonous. Distributed in Inner Mongolia, Northeast, and Southern China.

1014 *Clitocybe sinopica* (Fr.) P. Kumm.

Pileus 4–11 cm broad, oblate spherical at first, infundibuliform at maturity, brown red to ochraceous, white tomentulose, even at margin. Context white, unchanging when injured. Lamellae decurrent, close, white at first, becoming light yellow gradually at maturity, unequal. Stipe 4–9 cm long, 0.4–1 cm thick, cylindrical, concolourous with pileus, hollow. Basidiospores 8–9.5 × 6.5–5 μm, broadly elliptical to subovate, smooth, hyaline.

In summer to autumn, solitary or gregarious on ground in broad-leaved forests. Edible. Distributed in Northeast, Southern, and Northwest China.

1015 *Clitocybe squamulosa* (Pers.) P. Kumm.

Pileus 3–7 cm broad, plane with a incurved margin when young, then umbelliform or oblate hemispherical, droopy at margin, often undulate or crenate or ribbed at margin, densely squamulose or fibrillose, brown, dirty brown to cinnamon brown, becoming lighter when dried or old, darker at disc, lighter at margin. Context white or light brown, thin, soft when wet, solid and brittle when dried. Lamellae decurrent, white to light brown, close to subdistant, narrow to medially wide, sometimes furcate or interveined. Basidiospores 6–7.5 × 4 μm, elliptical, smooth, hyaline, inamyloid.

On leaf litter in coniferous forests. Distributed in Northeast China.

1016 *Clitocybula familia* (Peck) Singer

Pileus 2–3 cm broad, oblate hemispherical to plane, slightly depressed at centre, light grey, radially lacerate to form appressed fibrillose squamules. Context white, thin. Lamellae sinuate, white. Stipe 3–5 cm long, 3–5 mm thick, cylindrical, white to greyish white, with grey to light grey squamules. Basidiospores 4–5.5 × 3.5–5 μm, globose to subglobose, smooth, hyaline, amyloid.

In summer to autumn, on rotten woods in forests. Edible according records. Distributed in Northeast China.

1017 *Clitopilus crispus* Pat.

Pileus 2–7 cm broad, oblate hemispherical to flat, white to pinkish white, depressed to concave at centre, incurved at margin; margin with radial ridges and spiny fibrils, fringed. Context white. Lamellae 2–3 mm broad, decurrent, unequal, white at first, cream to pink at maturity. Stipe 2–6 cm long, 0.3–1 cm thick, white. Basidiospores 6–7.5 × 4.5–5.5 μm, oval, broadly elliptical to elliptical, with 9–11 longitudinal ridges, pinkish.

In spring to autumn, on roadside or ground in forests in tropical zone. Utilizability unknown. Distributed in Southern China.

1018 *Clitopilus prunulus* (Scop.) P. Kumm.

Pileus 3–10 cm broad, oblate hemispherical at first, becoming plane, slightly concave to shallowly discoid, pruinose to even, undulate and incurved at margin, dry, white, dirty white or light grey. Context thick, white, with a strong smell. Lamellae decurrent, subclose, unequal, white to pink. Stipe 3–8 cm long, 1.5 cm thick, subcylindrical, often eccentric, glabrous, white to light grey, solid. Basidiospores 9–12 × 4–6 μm, broadly elliptical or subfusiform (or near football), with 6 longitudinal ridges.

In spring to summer, scattered or gregarious on grassland near forests. Distributed in Central, Southern, and Northwest China.

1019 *Clitopilus ravus* W.Q. Deng & T.H. Li

Pileus 3–4 cm broad, convex at first, plane at maturity; tawny, grey or greyish brown to brown, with brown tomentum. Context thick, white, unchanging when injured. Lamellae decurrent, close, unequal, with abundant lamellulae, white at first, soon becoming pallid fleshy pink to pink. Stipe 3.5–4.5 cm long, 0.4–1.2 cm thick, cylindrical, slightly enlarged at base, partially concolourous with pileus or pallid, orange grey to light brown, tomentose, unobviously striate, with white mycelium at base. Basidiospores 11–13 × 5.5–6.5 μm, amygdaliform or subfusiform, longitudinally ridged, pinkish, slightly thick walled.

In summer to autumn, solitary or scattered on ground in broad-leaved forests. Distributed in Central and Southern China.

1020 *Collybia cookei* (Bres.) J.D. Arnold

Pileus 0.3–1 cm broad, hemispherical at first, becoming plane at maturity, white to dirty white, light grey at centre; margin incurved when young, plane at maturity. Context white, thin. Lamellae adnate to decurrent, distant, concolourous with pileus, unequal. Stipe 2–5 cm long, 0.3–1 mm thick, cylindrical, nearly white, slightly twisted, solid. Sclerotium 0.6 cm long, subglobose, yellowish brown. Basidiospores 3.9–5.2 × 2.6–3.3 μm, elliptical to lacrimal, smooth, hyaline, inamyloid.

In summer to autumn, solitary or gregarious on leaf litter. Distributed in Northeast China.

This species is similar to *C. tuberosa*, the hyphae of sclerotium in this species are globose, whereas the hyphae in *C. tuberosa* are oblongly elliptical.

1021 *Conchomyces verrucisporus* Overeem

Pileus 2–4 cm broad, flabelliform to flat semicircular, white, dry, finely tomentulose, with thicker tomentum and brownish at base, becoming yellowish white, yellow to fleshy coloured when dried; estriate at margin. Context thick, nearly white to concolourous with pileus. Lamellae adnate or subdecurrent, wide, very close, unequal. Stipe short, light brown to brown, or dark dirty white. Basidiospores 6–7.5 × 4.5–5 μm, elliptical to subglobose, verrucose-echinulate, hyaline, inamyloid.

Gregarious on rotten wood of broad-leaved trees. Distributed in Southern China.

1022 *Coniolepiota spongodes* (Berk. & Broome) Vellinga

Pileus 2–4 cm broad, conical to hemispherical at first, then convex to nearly plane, with fagacious pinkish purple pulverulent appendages, white to nearly white on the background without pulverucence. Context thin, white. Lamellae free, white, close. Stipe 2–4.5 cm long, 2.5–5 mm thick, clavate, enlarged at base, with pinkish purple pulverulent appendages, concolourous with pileus. Basidiospores 4.5–6.5 × 3–3.5 μm, long elliptical, smooth, yellowish brown to darker brown, weakly dextrinoid.

Solitary on ground in broad-leaved forests. Distributed in Southern China.

1023 *Connopus acervatus* (Fr.) K.W. Hughes et al.

≡ *Collybia acervata* (Fr.) P. Kumm.

Pileus 0.5–3 cm broad, convex when young, incurved at margin, becoming plane, gradually uplifted at margin, reddish brown to purplish brown, becoming lighter at maturity, glabrous, hygrophanous at margin. Context concolourous with pileus or lighter, unchanging when injured. Lamellae subfree, close, light yellow to light brown, unequal. Stipe 4–10 cm long, 2–3 cm thick, cylindrical, central, slightly curved, purplish brown. Basidiospores 5–6 × 2–2.5 μm, lacrimal to elliptical, smooth, hyaline.

In summer to autumn, clustered on ground or decayed wood in coniferous and broad-leaved forests. Edible. Distributed in Northeast and Central China.

1024 *Conocybe apala* (Fr.) Arnolds

Pileus 1–3 cm broad, broadly campanulate or umbelliform to campanulate, thin and brittle, yellowish white to light yellowish brown, darker at disc, nearly white to yellowish white and finely striate at margin, sticky. Context dirty white, thin. Lamellae adnate, close, narrow, unequal, dirty white at first, becoming rusty yellow. Stipe 5–8 cm long, 1–3 mm thick, white or greyish white, pulverulently granular, cylindrical, equal to slightly enlarged at base, hollow. Basidiospores 12–18 × 6–10 µm, elliptical to oval, smooth, rusty brown.

In summer to autumn, solitary or gregarious on ground of grassland, roadside or lawn near forests. Distributed in Southern and Central China.

1025 *Conocybe arrhenii* (Fr.) Kits van Wav.

Pileus 1.1–2.4 cm broad, conical to campanulate when young, broadly campanulate at maturity; yellowish brown to greyish brown, darker at centre, with white tomentum, radially striate from disc to margin, hygrophanous. Context thin, yellowish brown. Lamellae free to adnate, light yellowish brown to greyish brown. Stipe 2.5–4 cm long, 1.5–2 mm thick, cylindrical, squamulose, light yellow at upper portion, darker at lower portion when fresh. Annulus medial to superior, white or concolourous with pileus, fagacious. Basidiospores 14.5–17 × 7–9.5 µm, broadly elliptical, with an apical germ pore, smooth, yellowish brown.

In summer to autumn, scattered or gregarious on rotten woods in broad-leaved forests. Distributed in Northeast China.

1026 *Conocybe macrospora* (G.F. Atk.) Hauskn.

Pileus 1.2–2 cm broad, conical to campanulate; incurved at margin when young, glabrous, dirty brown to light yellow brown, becoming lighter when mature or dried, darker at apex, lighter at margin. Context light yellow brown to brown, with indistinct odour to odourless. Lamellae adnate, yellowish brown to rusty brown, subclose, unequal. Stipe 4–7 cm long, 1.5–4 mm thick, cylindrical, straight and glabrous, rusty brown, becoming darker from apex to base. Basidiospores 22 × 7.5–12.5 µm, long cylindrical to cylindrical, sooth, light rusty.

In summer to autumn, gregarious on grassland or rotten wood. Distributed in Northeast China.

1027 *Coprinellus disseminatus* (Pers.) J.E. Lange

Pileus 5–10 cm broad, oval to campanulate at first, plane at maturity, light brown to yellow brown, with white to brown granular to flocculent scales, long striate at margin. Context nearly white, thin. Lamellae white at first, becoming brown to nearly black, not deliquescing or only slowly deliquescing at maturity. Stipe 2–4 cm long, 1–2 mm thick, white to greyish white. Annulus abscent. Basidiospores 6.5–9.5 × 4–6 µm, elliptical to oval, smooth, light greyish brown, with an apical germ pore.

In summer to autumn, on roadside, rotten woods in forests or grassland. It has been recorded as edible when young, but is poisonous when old and is so small that it is not recommended to eat. Distributed widely in China.

1028 *Coprinellus micaceus* (Bull.) Vilgalys et al.

Pileus 2–4 cm broad, oval to campanulate at first, then plane, uplifted at margin at maturity, light yellow, brown, yellowish brown, reddish brown to ochraceous brown, becoming lighter to grey toward margin, hygrophanous, with white granular crystals when young which are crystals gradually disappearing at maturity, long striate at margin. Context nearly white to light ochraceous brown, thin, brittle. Lamellae beige yellow at first, changing to black, deliquescing at maturity. Stipe 3–8.5 cm long, 2–5 mm thick, cylindrical, subequal, sometimes enlarged to clavate or bulbous at base, white, white pruinose, then glabrescent, gradually becoming light yellow, brittle, hollow. Annulus abscent. Basidiospores 7–10 × 5–6 μm, elliptical, smooth, greyish brown to dark brown, with a slightly truncated apical germ-pore.

In spring to autumn, fasciculate or gregarious on ground in broad-leaved forests. Recorded as edible when young, however it is not recommended to eat. Distributed widely in China.

1029 *Coprinellus radians* (Desm.) Vilgalys et al.

Pileus 0.2–0.6 cm broad, 0.2–0.8 cm when young, 0.5–2.5 cm broad at maturity, globose to oval at first, then expanded and upcurved at margin, with white hairy scales, ochraceous brown, olivaceous grey at centre, white at margin, squamulose and striate, lacerated at maturity. Context thin, greyish brown at first. Lamellae sinuate and free, white when young, then becoming black gradually, distant, unequal, even at margin. Stipe 2–6.5 cm long, 1–4 mm thick, cylindrical, enlarged downwards, brittle and fragile, hollow, often with bristle-like mycelium on stipe base and substrate. Basidiospores 10–12 × 6–7.5 μm, elliptical, smooth, greyish brown to dark brown, with an obvious apical germ-pore.

In spring to autumn, gregarious on stumps or decayed woods. Distributed in Northeast and Northwest China.

1030 *Coprinopsis atramentaria* (Bull.) Redhead et al.

Pileus 3.5–8.5 cm broad, oval at first, then expanded to campanulate to conical, up-curling at margin at maturity, deliquescing to black "ink" when old, with brown scales, subglabrous at margin. Context thin, white at first, then becoming greyish white. Lamellae sinuate, close, unequal, white to greyish when young, then greyish brown to black, deliquescing to black "ink" when old. Stipe 3.5–8.5 cm long, 0.6–1.2 cm thick, cylindrical, enlarged downwards, white to greyish white, glabrous or fibrillosely squamulose, hollow. Basidiospores 7.5–10 × 5–6 μm, elliptical to broadly elliptical, smooth, dark greyish brown to blackish brown, with an obvious apical germ-pore.

In spring to autumn, fasciculate on decayed woods in forest, open field, roadside, countryside and park. Edible when young, poisonous at maturity, it is not recommended to eat. Distributed widely in China.

1031 *Coprinopsis cinerea* (Schaeff.) Redhead et al.

= *Coprinus radiatus* (Bolton) Gray
Pileus 2–5 cm broad, elliptical when young, white to light brown or greyish brown, with dirty white to argenteous tomentose scales when young, becoming obtuse conical and greyish brown to grey at maturity; margin recurved, radially lacerated, deliquescing from margin to disc. Context thin, white to pale grey. Lamellae close, free, white at first, then becoming greyish brown. Stipe 5–12 cm long, 0.4–1 cm thick, hollow, sparsely white tomentose, enlarged at base and then long radicant. Basidiospores 8.4–11.8 × 5.8–7.8 μm, oval to elliptical, smooth, dark brown.

In spring to autumn, solitary or scattered on dungy soil and hay piles. Distributed in Northeast, Northern, Central and Northwest China.

1032 *Coprinopsis friesii* (Quél.) P. Karst.

Pileus 0.3–0.8 cm broad, 0.2–0.6 cm high, 1.5 cm broad when fully expanded, conical, oval to elliptical at first, white, ochraceous at centre, with light ochraceous macules when universal veil lacerated. Context thick, greyish white. Lamellae free, white when young, then becoming grey to black gradually, subclose, unequal. Stipe 3–5 cm long, 1–1.5 mm thick, cylindrical, white to greyish white, with sparse flocculent scales, sometimes clavate at base. Basidiospores 6.1–9.7 × 5.3–7.3 μm, oval to rhomboid, rounded at apex, smooth, greyish brown to dark reddish brown, with an obvious apical germ-pore.

In spring to autumn, gregarious on grassland. Distributed in Northeast China.

1033 *Coprinopsis nivea* (Pers.) Redhead et al.

Pileus 2–3 cm broad, white, oval to campanulate, densely with white farinose granular velar remnants. Context white. Lamellae free, white at first, then becoming grey, nearly black at maturity. Stipe 7–10 cm long, 3–6 mm thick, white to dirty white, covered with farinose scales, gradually glabrescent. Annulus abscent. Basidia 25–35 × 12–15 μm. Basidiospores 12–16 × 10–14 × 7–9 μm, elliptical in side view, citriform in frontal view, smooth, nearly black, with an apical germ-pore.

In summer to autumn, on yak dung in alpine meadow. Distributed in Qinghai-Tibet region.

1034 *Coprinus comatus* (O.F. Müll.) Pers.

Pileus 6–11 cm high, 3–6 cm broad, cylindrical when young, then campanulate, plane finally, white, silky lustrous, at first, light earthy yellow at apex, glabrous, then becoming darker gradually; pileipellis lacerated into appressed and decurved scales; margin finely striate, sometimes pink. Context white, thick at centre, thin around. Lamellae white at first, then becoming pinkish grey to black, deliquescing when old to black "ink" together with pileal margin. Stipe 7–25 cm long, 1–2 cm thick, cylindrical, fusiform and forming a subterranean radicant base into soil, glabrous, white, hollow, enlarged at ground surface and then tapering downwards under the ground. Annulus white, membranous, then movable, fagacious. Basidiospores 12.5–19 × 7.5–11 μm, elliptical, smooth, black.

In summer to autumn, gregarious or solitary on grasslands, open field in forests, roadside or field. Edible when young, cultivable. Distributed in Northeast, Northern, and Central China.

1035 *Cortinarius alboviolaceus* (Pers.) Fr.

Pileus 3–10 cm broad, hemispherical at first, slightly prominent at centre, incurved at margin, then plane, dry, glabrous, white to weak violaceous at margin, pale yellowish brown at disc, with slightly violaceous tinct. Context light violaceous, thick, becoming darker when injured. Lamellae close, light violaceous at first, then changing to brown to ferruginous brown, unequal, sinuate, even to undulate at edge. Stipe 4–10 cm long, 1–2 cm thick, becoming thicker downwards, slightly enlarged at base, violaceous at upper portion, light violaceous to whitish violaceous at lower portion, changing to greyish violaceous when injured, solid. Annulus superior, cortinoid, whitish, greyish violaceous to light yellowish brown or concolourous with basidiospores, fagacious. Basidiospores 8–11 × 5–6.5 μm, elliptical to elliptical to long ellipsoid, rough, verruculose, ferruginous brown.

In summer to autumn, scattered on ground in mixed or coniferous forests. Edible, however also known to be poisonous in some records. Distributed in Northeast and Qinghai-Tibet Region, China.

1036 *Cortinarius armillatus* (Fr.) Fr.

Pileus 3–8 cm broad, hemispherical to campanulate at first, prominent at centre, incurved at margin, then plane, often appendiculate with light red remnants, brick red to reddish brown, reddish brown to deep brown at disc, even. Context light reddish brown to light brown, thinner. Lamellae adnate to sinuate, cinnamon brown at first, then changing to yellowish brown to ferruginous brown, distant, even at edge. Stipe 6–12 cm long, 0.5–1 cm thick, subcylindrical, often with several brick red annular zones at upper and middle portion, with pieces of fibrous remnants at lower portion, slightly enlarged at base, radicant, solid. Partial veil superior, cortinoid, tincted with ferruginous brown, fagacious. Basidiospores 8.5–12 × 4.5–7 μm, elliptical, verruculose, ferruginous brown.

In autumn, on ground in broad-leaved forests. Distributed in Northeast and Southern China.

1037 *Cortinarius bivelus* (Fr.) Fr.

Pileus 3.8–7.5 cm broad, hemispherical at first, then convex to plane gradually, sometimes slightly prominent at centre, yellowish brown to ferruginous brown, sometimes hygrophanous, darker at disc, with appressed fibrils, incurved to straight at margin, sometimes with white partial veil remnants. Context white to whitish. Lamellae adnate, light yellowish brown when young, changing to ferruginous brown at maturity, close, unequal. Stipe 3.5–7 cm long, 0.6–1.5 cm thick, subcylindrical, enlarged at base, with white woolly fibrils at first, changing to light brown to drab, solid. Partial veil cortinoid, white, then changing to ferruginous brown when covered with basidiospores, often leaving annular zones on stipe at maturity. Basidiospores 7.5–10.5 × 5–6.5 μm, elliptical, verruculose, light yellowish brown to yellowish brown.

In summer to autumn, solitary to gregarious on ground in broad-leaved forests. Distributed in Northeast, Northwest and Qinghai-Tibet Regions, China.

1038 ***Cortinarius bolaris*** (Pers.) Fr.

Pileus 2–3.5 cm broad, hemispherical at first, then convex to plane, sometimes prominent at centre, light yellow to brownish yellow, densely with appressed reddish brown tomentose squamules, incurved to plane at margin. Context white, changing to orange when injured. Lamellae sinuate, light yellowish brown to yellowish brown, with slightly olive tinge, close, unequal. Stipe 2.5–4.5 cm long, 3–6 mm thick, cylindrical, slightly enlarged at base; solid when young, hollow at maturity, light yellow, densely with appressed reddish brown tomentose scales. Partial veil cortinoid, whitish when young, then changing to reddish brown at maturity as being covered with basidiospores. Basidiospores 6–8 × 5–6 μm, subglobose to oval, verruculose, light brown.

In summer to autumn, solitary to gregarious on ground in broad-leaved or mixed forests. Poisonous according to foreign records. Distributed in Central China.

1039 ***Cortinarius* cf. *bovinus*** Fr.

Pileus 4–8 cm broad, conical to campanulate at first, incurved at margin, then plane, brown to chestnut brown, dirty brown to greyish brown when dried, with appressed fibrils. Context light brown, thick at centre, becoming thinner toward margin. Lamellae sinuate, wide, close, light brown at first, changing to dark brown to ferruginous brown, even and undulated at edge. Stipe 5–8 cm long, 1–1.6 cm thick, cylindrical, or tapering toward apex, bulbously enlarged at base, dirty brown, with dirty white tomentum, rarely with dirty white fagacious annular zones at upper portion, solid. Partial veil superior, cortinoid, white to with ferruginous brown tinge, fagacious. Basidiospores 7.5–11 × 5–6.5 μm, elliptical, verruculose, ferruginous brown.

In autumn, on ground in broad-leaved or mixed forests. Distributed in Northeast and Central China.

1040 ***Cortinarius* cf. *caperatus*** (Pers.) Fr.

≡ *Agaricus caperatus* Pers.
≡ *Rozites caperatus* (Pers.) P. Karst.

Fruiting body fleshy, medium to large. Pileus 4–8 cm broad, oblate hemispherical at first, then plane, prominent at centre, yellow to light yellowish brown, glabrous or occasionally with remnants of pulverulent universal veil, roughly rugulose when old and dried. Context white to light yellow, thick. Lamellae adnate or sinuate, creamy white at first, then changing to ferruginous brown, with white interveins, wide, close, even at edge, unequal. Stipe 8–14 cm long, 1.2–2 cm thick, cylindrical, light yellow to yellow, fibrous, solid, with remnants of yellow veil at base. Annulus superior to medial, membranous, yellowish white. Basidiospores 10–15 × 7–8.5 μm, elliptical, rough, verruculose, light ferruginous brown.

In autumn, on ground in broad-leaved or mixed forests. Distributed in Northeast China.

1041 *Cortinarius cinnamomeus* (L.) Fr.

Pileus 2–6 cm broad, obtuse or umbonate at centre, dry, greenish yellow, yellowish brown or reddish brown, dark at disc, densely with light yellowish brown squamules or radially fibrillose. Context light orange or straw yellow, changing to brown or ferruginous brown, thin, mild. Stipe 3–8 cm long, 3–7 mm thick, cylindrical, subequal or slightly curved, sometimes bulbous at base, yellow to earthy yellow, with brown hairs, changing to darker when injured, solid to hollow, often with yellow rhizomorphs at base. Partial veil superior, cortinoid, yellow, fagacious. Basidiospores 5.5–8.5 × 4–5.5 μm, limoniform to wide elliptical, rough, asperulate, light ferruginous brown.

In autumn, gregarious or subfasciculate on ground in coniferous forests with *Picea asperata*, etc. or in mixed forests. Edible. Distributed in Northeast, Northwest and Qinghai-Tibet Regions, China.

1042 *Cortinarius collinitus* (Pers.) Fr.

Pileus 4–10 cm broad, hemispherical to campanulate at first, incurved at margin, then plane with a prominent disc, viscid or glutinous when wet, glabrous, cinnamon, khaki to yellowish brown, then light cinnamon. Context white to light yellow, then yellow with yellowish brown, thick. Lamellae adnate or sinuate, close, light yellow to khaki at first, then changing to ferruginous brown, even at edge. Stipe 6–11 cm long, 0.8–1.5 cm thick, cylindrical, subequal or tapering downward, white to light yellow at upper portion, yellowish brown to brown at lower portion, covered with a slime layer at first, connected with cortina at upper edge, soft to solid. Partial veil superior, cortinoid, sometimes forming an inconspicuous annulus, white to tincted with ferruginous brown, fagacious. Basidiospores 10–14 × 5.5–7.5 μm, amygdaliform to elliptical, rough, verruculose, ferruginous brown to yellowish brown.

In autumn, on ground in mixed forests. Edible. Distributed in Northeast and Southern China.

The main characteristics of this species are the glutinous fruiting body, glabrous pileus, khaki to yellowish brown pileal colour, subequal or tapering stipe downwards, stipe base thin and pointed, white colour of upper stipe surface and yellowish brown to brown colour of lower stipe surface.

1043 *Cortinarius* cf. *croceicolor* Kauffman

Pileus 4–9 cm broad, hemispherical at first, incurved at margin, then plane, obtusely umbonate at centre, yellow to reddish yellow, densely brown squamulose or fibrillose at disc. Context white to light yellow, thick at centre, thinner at margin. Lamellae sinuate, close, wide, chrome yellow to yellowish brown, undulate at edge. Stipe 4–9 cm long, 0.5–1.3 cm thick, cylindrical, or enlarged downwards, light yellow to chrome yellow; bulbous at base, with chrome yellow rhizomorphs or universal veil, solid. Partial veil superior, cortinoid, white to ferruginous brown, fagacious. Basidiospores 6–9 × 5–7.5 μm, subglobose to wide elliptical, rough, verruculose, ferruginous brown to yellowish brown.

In autumn, on ground in broad-leaved or mixed forests. Distributed in Northeast China.

1044 *Cortinarius* cf. *elatior* Fr.

Pileus 3–9 cm broad, subglobose at first, then changing to conical, umbonate at centre, incurved at margin, then straight, glutinous, dirty yellow to yellowish brown at first, slightly olivaceous, khaki to dirty yellowish brown when dried, radially striate at margin, easily lacerated when dried. Context thick, whitish to light yellow. Lamellae sinuate or free, rarely adnate, close, wide, white to khaki, changing to ferruginous brown, interveined. Stipe 6–13 cm long, 0.6–1.2 cm thick, cylindrical, light purple when young, then light yellow, glutinous, cracked on lower portion, solid. Partial veil inconspicuous, fagacious, and almost leaving no vestige. Basidiospores 12.5–15 × 6–9 μm, elliptical, rough, obviously verruculose, ferruginous brown to yellowish brown.

In autumn, on ground in broad-leaved or mixed forests. Edible. Distributed in Northeast, Southern, Central and Qinghai-Tibet Region, China.

1045 *Cortinarius emodensis* Berk.

Pileus 6–15 cm broad, plane and broadly umbonate, yellowish brown, radially rugulose, grey to brown tomentose-squamose. Lamellae violet at first, then changing to light brown to brown. Stipe 10–15 cm long, 1–3.5 cm thick, cylindrical. Context yellow, unchanging when injured; violet at apex when young, changing to light yellow to light brown at maturity. Annulus superior, thick, membranous, whitish, light brown or concolourous with stipe. Basidiospores 13–16 × 8.5–11 μm, amygdaliform to elliptical, minutely verruculose, ferruginous brown to yellow brown..

In summer, on ground of alpine and subalpine forests. Edible. Distributed in Qinghai-Tibet Region, China.

1046 *Cortinarius emodensis* var. *vinacea* (Zhu L. Yang & W.K. Zheng) Peintner et al.

Pileus 6–15 cm broad, convex to plane, purple to violet, radially rugulose, grey to brown tomentose-squamose. Context whitish to yellowish brown. Lamellae violet at first, then changing to light brown to brown. Stipe 10–15 cm long, 1–3.5 cm thick, cylindrical, light violet at upper portion, light brown at lower portion. Annulus superior, thick, membranous, concolourous with stipe or slightly lighter. Basidiospores 13–15 × 9–10.5 μm, amygdaliform, elliptical, minutely verruculose, ferruginous brown.

In summer to autumn, on ground of alpine and subalpine forests. Edible. Distributed in Qinghai-Tibet Region, China.

1047 *Cortinarius galeroides* Hongo

Pileus 1–3 cm broad, conical at first, then bamboo-hat-shaped to plane, conspicuously umbonate at centre, ochraceous, darker at disc, lighter at margin, striate, hygrophanous. Context light yellow, thin. Lamellae light khaki, adnate to free, subdistant, unequal. Stipe 3–5 cm long, 0.4–1 cm thick, subcylindrical, tapering downwards, lighter than pileus, finely powdery-granular at apex, soft to solid insides. Partial veil superior, cortinoid, white at first, then changing to ferruginous brown, fagacious. Basidiospores 7.5–9.5 × 4.5–5.5 µm, elliptical, tuberculate, ferruginous brown.

In summer to autumn, gregarious on ground in mixed forests. Distributed in Northeast and Central China.

1048 *Cortinarius hemitrichus* (Pers.) Fr.

Pileus 3–5 cm broad, hemispherical to conical at first, sparsely with dark brown tomentum, expanding to bamboo-hat-shaped or broadly campanulate; epidermis easily peeled off, cracked when dried; surface brown to ferruginous brown, dark brown at disc, often with dirty white remnants at margin. Context light yellow, thin. Lamellae adnate to sinuate, light yellow brown at first, then ferruginous brown, unequal, close, wide. Stipe 12–14 cm long, 0.7–1.6 cm thick, cylindrical, dirty white to pale purplish brown at first, then yellowish brown to dirty brown, soft to solid insides. Partial veil superior, cortinoid, sometimes forming an annulus, white at first, then ferruginous brown. Basidiospores 5–10 × 4–6 µm, wide elliptical, rough, verruculose.

In summer to autumn, on ground in mixed forests. Edible. Distributed in Northeast, Northwest and Qinghai-Tibet Region, China.

1049 *Cortinarius korfii* T.Z. Wei & Y.J. Yao

Pileus 4.2–13.5 cm broad, subglobose to campanulate when young, then convex to plane gradually, slightly prominent at centre, reddish brown to purplish brown when young, light yellowish brown at margin, then khaki brown to reddish brown, slightly with olive tint, darker at disc, not viscid, with reddish brown to dark reddish brown squamules, radiately striate and grooved at margin. Context white to dirty white, thick. Lamellae subadnexed, light greyish purple when young, changing to light greyish purple gradually, khaki brown to dark ferruginous brown at maturity, tincted with bluish purple, close, unequal. Stipe 4.5–21 cm long, 0.9–2 cm thick, cylindrical, enlarged at base, white in upper portion, light yellowish when young in lower and middle portions, changing to light yellowish brown to light brown gradually, not viscid, with yellowish brown to dark brown fibrils, longitudinally striate. Partial veil cortinoid, whitish when young, then yellowish brown due to covering basidiospores. Basidiospores 10–16 × 8–10 µm, elliptical to amygdaliform, verruculose, yellowish brown to brown.

In summer to autumn, solitary to gregarious on ground in broad-leaved forests. Distributed in Central China.

1050 *Cortinarius multiformis* Fr.

Pileus 5–9 cm broad, hemispherical at first, slightly prominent or flat at centre, then plane, subviscid when wet, epidermis easily peeled off, light yellow to khaki, then light ferruginous brown to yellowish brown. Context white, thick. Lamellae adnate to sinuate, white to light khaki at first, changing to yellowish brown to nearly ferruginous when dried, unequal, close, wide, even at edge. Stipe 3–8 cm long, 0.8–2.5 cm thick, cylindrical or gradually thicker downwards, bulbous at base, with white fibrils, light yellow, then yellow, changing to yellow or yellowish brown at base when dried, soft insides. Partial veil superior, cortinoid, white, fagacious. Basidiospores 9–12.5 × 6–7.5 µm, elliptical to long elliptical, verruculose, ferruginous brown.

In summer to autumn, on ground in mixed or broad-leaved forests. Edible. Distributed in Northeast, Central, and Southern China.

1051 *Cortinarius* cf. *paleaceus* (Weinm.) Fr.

Pileus 2–5 cm broad, hemispherical to campanulate at first, prominent at centre, incurved at margin, then plane, dark brown to greyish brown, with greyish brown fibrils. Context dirty white, thin. Lamellae sinuate to subfree, light cinnamon when young, dark cinnamon to ferruginous brown at maturity, close, undulate or serrate at edge. Stipe 3–7 cm long, 0.6–1.2 cm thick, clavate, gradually thicker downwards, enlarged at base, dirty white to light brown at first, then dirty brown, soft to hollow. Partial veil superior, cortinoid, white, fagacious. Basidiospores 6.5–8.5 × 4–5 µm, long elliptical, rough, verruculose, ferruginous brown.

In autumn, on ground in mixed forests. Distributed in Northeast China.

1052 *Cortinarius pholideus* (Lilj.) Fr.

Pileus 3–5 cm broad, campanulate at first, incurved at margin, then plane, umbonate at centre, dark brown to cinnamon from inside to outside; with densely light brown scales. Context thin, white with violetish tint at first, then changing to dirty white. Lamellae adnate to sinuate, close, wide, white with violaceous tint at first, then khaki to light brown, even at edge. Stipe 4–9 cm long, 0.5–1 cm thick, cylindrical, enlarged at base, purplish in upper and middle portions, dirty brown in lower portion, with numerous brown scales, solid. Partial veil superior, cortinoid, white, fagacious, leaving a membranous annulus on stipe. Basidiospores 6.5–8 × 4.5–6 µm, elliptical, rough, verruculose, ferruginous brown.

In summer to autumn, on ground in broad-leaved or mixed forests. Distributed in Qinghai-Tibet Region, and Northeast China.

1053 *Cortinarius pseudosalor* J.E. Lange

Pileus 3–9 cm broad, campanulate at first, or prominent to conical at disc, then plane, viscid, ochraceous yellow to yellowish brown, dark brown at centre, undulately striate at margin. Context light yellowish brown, thin. Lamellae sinuate, thin, distant, brown to yellowish brown at first, then light cinnamon, undulate at edge. Stipe 10–13 cm long, 0.6–1.5 cm thick, cylindrical, white with purplish tint at first, gradually fading, changing to yellowish white to yellowish brown, viscid, soft to hollow. Partial veil superior, cortinoid, white at first, then ferruginous brown, with a fagacious annulus. Basidiospores 11.5–16 × 7.5–10 μm, elliptical, rough, verruculose, ferruginous brown.

In autumn, on ground in mixed forests. Distributed in Northwest and Northeast China.

1054 *Cortinarius purpurascens* Fr.

Pileus 4–11 cm broad, campanulate at first, incurved at margin, then plane, slightly prominent at centre, viscid when wet, violet, then changing to khaki to light brown, persistently purple at margin. Context thick, light purple at first, then bluish violet. Lamellae sinuate, close, wide, bluish violet at first, changing to dark purple when injured, changing to khaki to ferruginous brown when dried or at age. Stipe 4–12 cm long, 1.2 cm thick, cylindrical, enlarged at base, often bulbose, light purple to dirty white, often with purple fibrils, solid. Partial veil superior, cortinoid, bluish purple, fagacious. Basidiospores 8.5–10.5 × 4.5–5.5 μm, oval to elliptical, verruculose, ferruginous brown.

In autumn, on ground in mixed or broad-leaved forests. Distributed in Northeast, Central, and Southern China.

1055 *Cortinarius rufoolivaceus* (Pers.) Fr.

Pileus 5–10 cm broad, hemispherical at first, then convex to plane, reddish brown to purplish brown at centre, becoming lighter gradually outwards, reddish brown to light brown, lustrous, viscid when wet, keeping involved at margin for a long time. Context white, slightly purplish, thick. Lamellae sinuate, olive when young, yellowish brown when mature, close, unequal. Stipe 4–6 cm long, 1–2 cm thick, central, cylindrical, enlarged at base, solid, light purple, purplish brown at base, not viscid. Obviously deep purple when dried. Partial veil cortinoid, attached to medial part of stipe for a long time, ferruginous brown. Basidiospores 9.5–13 × 6–7.5 μm, amygdaliform, obviously verruculose, brown to reddish brown.

In summer to autumn, gregarious on ground in mixed forests. Distributed in Northwest and Qinghai-Tibet Region, China.

1056 *Cortinarius sanguineus* (Wulfen) Fr.

Pileus 2–6 cm broad, oblate hemispherical at first, slightly prominent at centre, then plane and slightly depressed, sanguineous to purplish brown, becoming lighter when dried, with tomentose squamules when young, then glabrescent. Context light sanguineous to dark sanguineous, thin. Lamellae adnate, close, wide, even at edge. Stipe 4–9 cm long, 4–7 cm thick, equal, twisted, sparsely fibrillose, sanguineous, becoming darker when injured, fibrous, hollow. Partial veil superior, cortinoid, sanguineous, fagacious. Basidiospores 6.5–9 × 4–6 μm, long elliptical, rough, verruculose, ferruginous brown.

In summer to autumn, on ground in coniferous or mixed forests. Edible. Distributed in Northwest and Qinghai-Tibet Region, China.

1057 *Cortinarius semissangnineus* (Fr.) Gillet

Pileus 3–6 cm broad, campanulate at first, becoming plane gradually, obtuse or prominent at centre, yellow to yellowish brown, with brown fibrils occasionally. Context thin, hard and brittle, dirty white to light yellow. Lamellae adnate, then sinuate, close, narrow, sanguineous to dark red, then yellowish brown to ferruginous brown. Stipe 4–7 cm long, 4–7 cm thick, subcylindrical, chrome yellow to orange yellow, with yellowish brown fibrils, fibrous, solid. Partial veil superior, cortinoid, ferruginous brown, fagacious to partially attached to upper portion of stipe. Basidiospores 6–7 × 3.5–4.5 μm, elliptical, asperulate, ferruginous brown.

In summer to autumn, on ground in mixed forests. Distributed in Northeast China.

1058 *Cortinarius subbalaustinus* Rob. Henry

Pileus 5.2–7.2 cm broad, plane, sometimes centrally prominent slightly, orange brown to reddish brown, hygrophanous when wet, glabrous, straight and lighter at margin. Context yellowish brown to light reddish brown. Lamellae sinuate, light brown when young, then changing to ferruginous brown at maturity, close, unequal. Stipe 5.3–8.5 cm long, 0.9–8.5 cm thick, cylindrical, sometimes slightly enlarged at base, light brown, often with white fine fibrils, solid. Partial veil white, then changing to ferruginous brown as being covered with basidiospores. Basidiospores 6–9.5 × 4–5.5 μm, elliptical, verruculose, yellowish brown to brown.

In summer to autumn, solitary to gregarious on ground in mixed forests. Distributed in Central and Qinghai-Tibet Region, China.

1059 *Cortinarius subpurpurascens* (Batsch) Fr.

Pileus 6–11 cm broad, hemispherical at first, incurved at margin, then expanded to plane, often depressed at centre, undulate and shallowly grooved-striate at margin, light yellow to light khaki, typically with purplish tint, subviscid when wet, even when dried. Context white, thick. Lamellae adnate, close, wide, light cinnamon to slightly purplish when young, then changing to light brown at maturity, even at edge. Stipe 5–9 cm long, 1–1.5 cm thick, mostly clavate, gradually thicker downwards, enlarged at base, white in upper portion, whitish violet from lower portion to base, becoming deeper violet downwards, occasionally with yellowish brown tomentose squamules, soft to solid insides. Partial veil cortinoid to fibrillose; often ferruginous brown, inconspicuous in later stage. Basidiospores 8.5–10.5 × 5–6.5 μm, elliptical, verruculose, light ferruginous.

In autumn, on ground in mixed or broad-leaved forests. Distributed in Northeast China.

1060 *Cortinarius subtorvus* Lamoure

Pileus 4.5–6.6 cm broad, hemispherical to convex when young, then plane gradually, often slightly prominent at centre, dark ferruginous brown to dark purplish brown when wet, slightly subhygrophanous, glabrous, fibrillose, incurved to decurved at margin, append**iculate** with white partial veil when young. Context white, thick. Lamellae subsinuate, light purplish brown when young, then changing to dark ferruginous brown at maturity, close, unequal. Stipe 4.3–6.5 cm long, 0.7–1.3 cm thick, central, subcylindrical, enlarged at base, with white woolly fibrils, often slightly purple, light brown in middle and lower portions, usually annular zonate, solid. Partial veil white, then changing to ferruginous brown as being covered with basidiospores. Basidiospores 7.5–10 × 4.5–5.5 μm, elliptical to amygdaliform, verruculose prominent, light yellowish brown to yellowish brown.

In summer to autumn, on ground under bushes, or on ground in forests according to foreign reports. Distributed in Central and Qinghai-Tibet Region, China.

1061 *Cortinarius tenuipes* (Hongo) Hongo

Pileus 3–5 cm broad, hemispherical, convex to plane with a prominent centre, yellow to yellowish brown. Context dirty white to beige, unchanging when injured. Lamellae white at first, then changing to light brown to yellowish brown. Stipe 5–10 cm long, 0.7–1 cm thick, cylindrical, white to beige, with yellowish brown filamentous scales. Partial veil superior, forming an annulus, cortinoid, fagacious. Basidiospores 6.5–8 × 3.5–4.5 μm, subamygdaliform, inconspicuously finely asperulate, subglabrous, ferruginous brown.

In summer to autumn, on ground in mixed or broad-leaved forests. Edible. Distributed widely in China.

1062 *Cortinarius traganus* (Fr.) Fr.

Pileus 5–11 cm broad, hemispherical at first, then plan to flat, slightly prominent at centre; dirty white with light violaceous tinge which fading out to paler at maturity, becoming dirty white to light yellowish brown, with yellowish brown tomentum, appendiculate with flocculent remnants at margin. Context light yellowish brown, thick. Lamellae adnate, then sinuate, subclose, narrow, ferruginous brown to yellowish brown. Stipe 5–11 cm long, 1–2 cm thick, subclavate, enlarged at base, dirty white to yellowish brown, hairy or fibrillose-tomentose, solid. Partial veil superior, cortinoid, light violaceous, fagacious. Basidiospores 8–11 × 4.5–6.5 μm, elliptical, verruculose, ferruginous brown.

In summer to autumn, on ground in coniferous forests. Distributed in Central and Northeast China.

1063 *Cortinarius trivialis* J.E. Lange

Pileus 3–8 cm broad, conical to hemispherical when young, then becoming convex to plane gradually, often with a low umbo at centre; light yellowish brown, brown to reddish brown, sometimes with slightly olivaceous tint, lustrous, lubricous when wet, often remaining incurved at margin for a long time. Context white to whitish, thin. Lamellae subadnexed, cream when young, slightly with bluish grey tinge, ferruginous brown at maturity, close, unequal. Stipe 4–10 cm long, 0.5–1 cm thick, cylindrical, sometimes tapering as a radicant base rooting into ground, wrapped with a fibrillose membranoid sheath in middle and lower portions, whitish to yellowish brown, viscid when wet, often with subparallel annular zones. Basidiospores 10–15 × 6.5–8 μm, amygdaliform, conspicuously verruculose, yellowish brown.

In summer to autumn, solitary to gregarious on ground in broad-leaved, coniferous and mixed forests. Distributed in Northeast, Northern, Northwest, and Qinghai-Tibet Region, China.

1064 *Cortinarius* cf. *venetus* (Fr.) Fr.

Pileus 3–6 cm broad, hemispherical when young, becoming convex to plane at maturity, often prominent at centre; greenish brown at disc, lighter at margin, with appressed fibrillose scales. Context light yellowish green, thin. Lamellae sinuate to subadnate, yellowish green when young, ferruginous brown at maturity, close, unequal. Stipe 3–6.5 cm long, 0.5–1 cm thick, cylindrical, enlarged or with a rooting pseudorhiza at base, solid when young, then hollow at maturity; light yellowish green to light green brown. Partial veil cortinoid, slightly tinged with yellowish green. Basidiospores 5.5–7 × 5–6 µm, subglobose, verruculose, yellowish brown.

In summer to autumn, solitary to gregarious on ground in broad-leaved or mixed forests. Distributed in Central and Qinghai-Tibet Region, China.

1065 *Crepidotus applanatus* (Pers.) P. Kumm.

Pileus 1–4 cm broad, flabelliform, subsemicircular or reniform, flat, glabrous, hygrophanous when wet, white or yellowish white, often partially covered with brown basidiospores, becoming brownish or light khaki, white to yellowish or light pinkish yellow when dried; margin striate when wet, thin, incurved; base with white soft hairs. Context thin, white to dirty white, soft. Lamellae radiating from an attachment base, decurrent, close, unequal, white at first, then changing to light brown or cinnamon. Stipe absent or short. Basidiospores 4.5–7 × 4.5–6.5 µm, wide elliptical, globose to subglobose, densely spinulose, or asperulate to minutely spinulose-verruculose, light brown or ferruginous.

In summer to autumn, gregarious, superposed or subimbricate on rotten woods of broad-leaved trees. Distributed in Northeast China.

1066 *Crepidotus badiofloccosus* S. Imai

Pileus 1–5.5 cm broad, subflabelliform, conchate or subhemispherical, incurved at margin, yellowish white to dirty whitish yellow, densely with brown or dark brown hairy squamules, densely with yellowish brown or yellowish white soft hairs at base. Context thicker at base, white. Lamellae yellowish white to dirty yellowish white, becoming brownish yellow to greyish brown. Stipe absent or nearly so. Basidiospores 5–7 × 5.5–6.5 μm, globose, with a minute apiculus, minutely verruculose, yellowish brown.

In summer to autumn, gregarious on twigs or rotten woods of broad-leaved trees. Edibility unknown. Distributed in Northeast, Northern and Northwest China.

This species is similar to *C. fulvotomentosus*. However, the pileus of the latter species is reniform or semicircular, light ferruginous, densely with squamules of withered leaf colour; and the basidiospores are 7–10 × 5–6.5 μm, subglobose, minutely spinulose or asperulate, and light ferruginous.

1067 *Crepidotus cinnabarinus* Peck

Pileus 1–2 cm broad, flabelliform to conchate, vermilion or scarlet, with minute hairy squamules, incurved at margin. Context thin, dirty white to pink. Lamellae beige to pink, vermilion at edge. Stipe lateral and short, whitish. Basidiospores 6.5–9 × 4.5–6 μm, wide elliptical to subglobose, verruculose, light ferruginous.

In summer to autumn, on rotten woods. Distributed in Central China.

1068 *Crepidotus mollis* (Schaeff.) Staude

Pileus 1–6 cm broad, semicircular, flabelliform, conchate, or campanulate at first, then convex to plane, translucent when immersed with water, viscid, wholly white to greyish white, or yellowish brown to brown when dried, slightly tincted with khaki, tomentose and greyish white pruinose, becoming glabrous when tomentum and pruinescence **evanescent**. Stipe absent. Context thin, gelatinous beneath cuticle, whitish. Lamellae decurrent to free, close, radiating from an attachment base, unequal; white at first, then brown, dark cinnamon or light ferruginous. Basidiospores 7.5–10 × 5–6.5 μm, elliptical to oval, smooth, light ferruginous.

In summer to autumn, gregarious or superposed on rotten woods. Edible. Distributed widely in China.

This species is similar to *C. calolepis*, however the pileus in the latter species has dense brown tomentose when young.

1069 *Crepidotus sulphurinus* Imazeki & Toki

Pileus 0.5–1 cm broad, flabelliform to conchate, yellow, dirty yellow to sulfur yellow, with minute hairy squamules at base, undulate or decurved at margin. Context thin, yellow. Lamellae distant, yellowish brown to ferruginous brown. Stipe lateral and short. Basidiospores 9–10 × 8–9 μm, globose to subglobose, verruculose, light ferruginous.

In summer to autumn, on rotten woods. Distributed in Central and Southern China.

1070 *Crepidotus uber* (Berk. & M.A. Curtis) Sacc.

Pileus 0.5–1.2 cm broad, flabelliform, conchate, reniform to circular; membranous to fleshy; subviscid when wet, yellow to dark brown, glabrous, with white tomentose mycelium at base. Lamellae decurrent, radiating from a white tomentose base, close, ochraceous to ferruginous or brown when old. Context thin, white. Stipe absent or rudimentary, lateral. Basidiospores 6–8 × 4.5–5 μm, elliptical, smooth, thick-walled, light ferruginous.

In summer to autumn, gregarious or superposed on rotten woods of broad-leaved trees (*Quercus* sp.). Distributed in Central China.

1071 *Crinipellis scabella* (Alb. & Schwein.) Murrill

Pileus 0.3–1.5 cm broad, convex at first, becoming hemispherical gradually, with radial brown to reddish brown fibrils, becoming darker toward disc. Context white, unchanging when injured. Lamellae free to adnate, distant, unequal, white, even at edge. Stipe 0.5–3 cm long, 1–1.5 mm thick, cylindrical, brown, tomentulose. Basidiospores 8.5–9.5 × 4.5–6 μm, wide elliptical to long elliptical, smooth, hyaline, inamyloid.

In summer to autumn, clustered or scattered on rotten woods of broad-leaved trees. Distributed in Northeast China.

1072 *Cuphophyllus pratensis* (Fr.) Bon

Pileus 4–7 cm broad, convex to hemispherical at first, then plane, lacerated at margin, light apricot yellow to orange, glabrous; margin even when young, becoming undulate gradually. Context white, unchanging when injured. Lamellae subdecurrent, distant, light apricot yellow to creamy yellow, unequal, subeven at edge. Stipe 2.5–7 cm long, 1–2 cm thick, cylindrical, light apricot yellow to creamy yellow, slightly striate. Basidiospores 5–7.5 x 4–5 μm, elliptical, smooth, hyaline, inamyloid.

In summer to autumn, scattered on ground in mixed or coniferous forests. Edible. Distributed in Northeast China.

1073 *Cyptotrama asprata* (Berk.) Redhead & Ginns

Pileus 2–3 cm broad, hemispherical to plane, orange red, yellow to light yellow, with orange red to orange pyramidal squamules, incurved at margin. Context thin, dirty white to light yellow. Lamellae adnate, unequal, white. Stipe 2–4 cm long, 2.5–4 mm thick, cylindrical, whitish to beige, with yellow to light yellow squamules. Basidiospores 7–9 x 5–6.5 μm, subamygdaliform, smooth, hyaline, inamyloid.

In summer to autumn, on rotten woods. Distributed in Southern and Central China.

1074 *Cystoderma amianthinum* (Scop.) Fayod

Pileus 2–4 cm broad, plane, with a small prominence at disc, yellow to yellowish brown, concolourously verruculose, uneven. Context white. Lamellae white to beige. Stipe 5–7 cm long, 3–6 mm thick, cylindrical, white to beige and glabrous above annulus, densely with small brownish yellow squamules below annulus. Annulus fagacious. Basidiospores 5–6.5 x 2.5–3.5 μm, smooth, hyaline, amyloid.

In summer to autumn, on ground in mixed forests. Distributed widely in China.

1075 *Cystoderma carcharias* (Pers.) Fayod

Pileus 2–5 cm broad, conical when young, then broadly conical, campanulate to plane at maturity, minutely granular or pulverulent; dirty white to light grey, often with pinkish brown tint, often appendiculate with velar remnants at margin. Context white, thick, with a pungent odour. Lamellae white to cream, narrow, with a decurrent tooth to stipe, even at edge. Stipe 2.5–5 cm long, 3–5 mm thick, cylindrical, sometimes enlarged at base, solid when young, hollow at maturity, glabrous and white above annulus, pulverulent or granular and white to cream below annulus. Annulus membranous, infundibuliform. Basidiospores 4–6 x 3–4 μm, elliptical, smooth, hyaline.

Solitary amongst mosses in forests. Distributed in Northeast China.

1076 *Cystoderma fallax* A.H. Sm. & Singer

Pileus 2–5 cm broad, convex at first, then becoming plane, yellowish brown to brown, densely granular. Context light yellowish brown, unchanging when injured. Lamellae decurrent, close, whitish to cream yellow, unequal, subeven at edge. Stipe 3–6 cm long, 3–6 mm thick, central, light yellowish brown to brown, granular below annulus, white and not granular above annulus. Annulus superior to medial, concolourous with stipe, fagacious. Basidiospores 3.5–5 × 2.8–4 µm, elliptical, smooth, hyaline, amyloid.

In summer to autumn, on ground in mixed forests. Edible. Distributed in Northeast China.

1077 *Cystoderma granulosum* (Batsch) Fayod

Pileus 2–4 cm broad, plane, prominent at centre, brown to reddish brown, with minute concolourous warts or verrucae, uneven. Context white. Lamellae white to beige. Stipe 5–7 cm long, 3–6 mm thick, cylindrical, reddish flesh, glabrous above annulus, with densely brown to reddish brown squamules below annulus. Annulus fagacious. Basidiospores 3.5–4.5 × 2.5–3 µm, elliptical, smooth, hyaline, inamyloid.

Summer to autumn, on ground in mixed forests or grass. Distributed widely in China.

1078 *Cystolepiota seminuda* (Lasch) Bon

Pileus 0.5–2 cm broad, white to beige, beige to pale yellow brown at centre, with white, light pink to light brown pulverulent squamules. Context white. Lamellae free, whitish to beige. Stipe 1.5–4 cm long, 1–2 mm thick, cylindrical, with white, pinkish to light brown pulverulent squamules, white to whitish at upper portion, reddish brown only at base, becoming light brown at lower portion, pinkish brown or wine-red when old, only white to whitish at apex; stipe context mostly light purplish red, whitish at apex. Annulus superior, white, fagacious. Basidiospores 3.5–4.5 × 2.5–3 µm, elliptical, smooth or inconspicuously verruculose, hyaline.

Summer to autumn, on humus in coniferous and broad-leaved forests. Distributed widely in China.

1079 *Delicatula integrella* (Pers.) Fayod

Pileus 0.8–1.5 cm broad, convex to plane, glabrous, white, radially translucently striate when wet. Context membranous, soft, white. Lamellae white, narrow, ridged, branched irregularly or in middle, distant, subdecurrent or notched at stipe, smooth at edge. Stipe 1–2.5 cm long, 0.5–1 mm thick, white, hyaline, glabrous or finely fibrillose, slightly enlarged and with white hairy mycelium at base, brittle. Basidiospores 5.5–9.5 × 3–5.5 µm, elliptical, smooth, hyaline.

In summer to autumn, gregarious on rotten woods. Distributed in Northwest and Northeast China.

1080 *Descolea flavoannulata* (Lj.N. Vassiljeva) E. Horak

Pileus 6–8 cm broad, light yellow, yellowish brown to dark brown, with small yellow squamules, radiately striate at margin. Context concolourous with pileus, unchanging when injured or slightly darker. Lamellae yellow at first, then changing to brown to ferruginous brown. Stipe 5–10 cm long, 0.5–2 cm thick, cylindrical, light yellow to yellowish brown, with velar remnants at base. Annulus superior, membranous, yellow. Basidiospores 13–16 × 7.5–9 μm, limoniform to amygdaliform, verruculose, ferruginous brown.

In summer to autumn, on ground in forests. Edible. Distributed in Central and Northeast China.

1081 *Entoloma abortivum* (Berk. & M.A. Curtis) Donk

Pileus 3–10 cm broad, oblate hemispherical at first, then plane, slightly depressed at centre, dry, greyish white, grey, light greyish yellow brown to light yellowish brown, even or undulate at margin. Context thick, white. Lamellae adnate to decurrent, close, thin, greyish white at first, becoming pink, unequal. Stipe 3–8 cm long, 0.5–1.5 cm thick, subcylindrical, solid, inconspicuously longitudinally striate, concolourous with pileus or lighter, with mycelium at base. Basidiospores 8–10.5 × 5.5–7 μm, angular, 5–6 sided, heterodiametric, pinkish.

In autumn, solitary, fasciculate or gregarious on ground in mixed forests with *Pinus* or *Quercus* trees, or in broad-leaved forests. Edible. Distributed in Central and Northeast China.

This species is similar to *E. sinuatum*, however the lamellae of the latter species is adnate or sinuate, attention should be paid when collecting for eating.

1082 *Entoloma alpinum* Xiao Lan He et al.

Pileus 0.9–3 cm broad, convex to plane when young, subinfundibuliform at maturity, depressed at centre; straw-coloured, yellow tomentose, slightly hygrophanous, estriate at margin. Context thin, white to concolourous with pileus or lighter. Lamellae adnate to decurrent, subdistant, white at first, then changing to pink. Stipe 0.9–1.8 cm long, 1–4 mm thick, central, cylindrical, concolourous with pileus or lighter, white fibrillose, white tomentose at base. With a weak farinaceous taste. Basidiospores 11.5–15 × 9–11.5 μm, isodiametric to heterodiametric, 6–10 sided, thick-walled, pinkish.

In summer, on ground in broad-leaved forests. Distributed in Central China.

1083 *Entoloma altissimum* (Mass.) E. Horak

Pileus 3.5–5 cm broad, conical to convex, dry, blue to blue purple, becoming green when injured, densely with brown tomentum especially at centre, striate. Context dark blue, becoming green when injured, thin, spicy. Lamellae 10–22 per cm at pileal margin, distant, wide, unequal, sinuate, dark blue, pinkish and with basidiospore colour at maturity, becoming green when injured. Stipe 5–7 cm long, 3–5 mm thick, cylindrical, light blue to blue, brown tomentose and striate, cartilaginous, hollow. Basidiospores 7.7–10 μm broad, cuboid, apiculate, pinkish.

In summer to autumn, solitary on ground in broad-leaved forests. Distributed in Southern China.

1084 *Entoloma angustispermum* Noordel. & O.V. Morozova

Pileus 1.5–2 cm broad, hemispherical when young, plane at maturity, depressed at centre, white to beige or slightly greyish yellow, slightly darker to light greyish brown in depressed disc, not viscid, striate and inconspicuously squamulose. Context white, thin. Lamellae up to 2 mm broad, adnate or sinuate, white at first, then pink. Stipe 5–7 cm long, 1–2 mm thick, cylindrical, hollow, brittle, white, translucent, glabrous, with white mycelium at base. Basidiospores 9–13 × 5.5–7.6 μm, heterodiametric, 6–8 sided, unconspicuous angular, thin-walled, pinkish.

On ground in broad-leaved and mixed forests. Distributed in Northeast China.

1085 *Entoloma azureosquamulosum* Xiao Lan He & T.H. Li

Pileus 1–6 cm broad, hemispherical, convex, then plane, estriate, densely with granular squamules, dark blue to with purplish blue, darker to nearly bluish black at centre. Context 2 mm thick at stipe, white tincted with blue of pileus colour. Lamellae up to 5 mm broad, sinuate or adnate, with a short decurrent tooth, close, thick, white at first, then pink, unequal. Stipe 4–5 cm long, 4–8 mm thick, cylindrical or subclavate, very brittle, concolourous with pileus or lighter, with dark blue granular-squamules, with white mycelium at base. Basidiospores 9–10.5 × 6.5–8 μm, heterodiametric, 5–7 sided, thick-walled, light pink.

Scattered on ground in broad-leaved forests. Distributed in Central and Southern China.

1086 *Entoloma bisporigerum* (P.D. Orton) Noordel.

Pileus 2.5–4 cm broad, plane, slightly depressed at centre, with a wide prominence at centre in depressed disc, greyish brown to light brown, often with fleshy pink, not viscid, hygrophanous, striate at margin. Context grey, thin. Lamellae up to 4.5 mm broad, adnate or sinuate, distant, thin, white to greyish white when young, light greyish brown with pink at maturity, with 2 tiers of lamellulae. Stipe 2.5–6.3 cm long, 2.5–4.5 mm thick, cylindrical, hollow, concolourous with pileus or light greyish brown, longitudinally striate when old. Basidiospores 11.5–13.5 × 7.5–9 μm, heterodiametric, 6–9 sided, obviously and irregularly angular, thick-walled, pinkish.

Scattered on ground in broad-leaved or mixed forests. Distributed in Northeast China.

1087 *Entoloma byssisedum* var. *microsporum* Esteve-Rav. & Noordel.

Pileus 0.5–2.5 cm broad, crepidotoid or pleurotoid, white, grey to greyish brown, tomentose. Context thin, white. Lamellae up to 2 mm broad, adnate, distant, whitish to pinkish, undulate at edge. Stipe 3–8 mm long, 1–3 mm thick, eccentric to lateral, short, with white mycelium at base. Basidiospores 7.5–9.5 × 5.5–7.5 μm, heterodiametric, 5–6 sided, distinctly angular, pinkish.

On rotten wood of broad-leaved trees. Distributed in Northeast, Northern and Central China.

1088 *Entoloma caeruleoflavum* T.H. Li & Xiao Lan He

Pileus 3–6 cm broad, convex, becoming plane at maturity, inconspicuously umbonate, with minute furfuraceous appendages, bluish black to nearly black at centre, lighter at margin, with bluish green, yellow and brown tinges, often tincted with pink colour of basidiospores, inconspicuously sulcate or striate. Context white, thin. Lamellae sinuate, close, bright yellow with pink, becoming pink to light reddish brown when injured, undulate at edge. Stipe 5–10 cm long, 0.4–1 cm thick, cylindrical, gradually thicker downward, blue to dark blue, longitudinally and slightly twistedly fibrillose striate, with white mycelium at base. Basidiospores 6.5–7.5 × 6.3–7.3 μm, isodiametric, 5–7 sided, pinkish.

Scattered on ground in mixed forests. Distributed in Central China.

1089 *Entoloma caesiellum* Noordel. & Wölfel

Pileus 1.5–3.5 cm broad, hemispherical to convex, then plane, depressed at centre, light yellowish grey, darker at centre, slightly light blue at margin, slightly hygrophanous, without conspicuous striae at margin, with denser light brown squamules at centre, with light brown tomentum or glabrous at margin. Context thin, whitish. Lamellae up to 5 mm broad, adnate, distant, white at first, then pink. Stipe 5.5–7.7 cm long, 1.5–3 mm thick, cylindrical, hollow, azure blue to blue grey, glabrous, with white mycelium at base. Basidiospores 9–10.5 × 6.5–8 μm, heterodiametric, 5–7 sided, thick-walled, pinkish.

In spring to autumn, scattered on ground in mixed forests. Distributed in Northeast China.

1090 *Entoloma caespitosum* W.M. Zhang

= Entoloma carneobrunneum W.M. Zhang

Pileus 3–5 cm broad, bamboo-hat shaped, convex-umbonate to plano-umbonate, conspicuously papillate at centre, light purplish red, pinkish brown to reddish brown, greyish brown at and around central umbo, glabrous; margin estriate, reflexed when mature. Context 0.5–1 mm thick at stipe, pinkish to light purplish red, odourless. Lamellae 2–5 mm broad, sinuate or adnate, 18–21 per cm at margin, unequal, white at first, then pink, light brown to brown when dried. Stipe 3–9 cm long, 2–6 mm thick, cylindrical, white to whitish, hollow, cartilaginous, with white mycelium at base. Basidiospores 8.5–10.5 × 6–7.5 μm, heterodiametric, 6–8 sided, subelliptical, apiculate, pink.

Cespitose or fasciculate on ground in broad-leaved forests. Distributed in Southern China.

1091 *Entoloma changchunense* Xino Lan He & T.H. Li

Pileus 3–15 mm broad, bluntly conical to convex at first, expanding to plano-convex or plane, slightly sulcate at margin, dark green-brownish and pale reddish brown, inconspicuously zonate, with dense suberect to repent hirsute-fibrils or squamulae. Lamellae emarginate with a short decurrent tooth, ventricose, subdistant, up to 3 mm broad, thick, grayish, finely eroded and concolorous at edge, unequal. Stipe central, 12–45 mm long, 0.5–2 mm thick, cylindrical, equal, densely with dark reddish or ferruginous hirsute-fibrils on dark greenish brown background, hollow, dry, with strigose hairs at base. Context thin, firm, concolorous with pileus, with mild odour and taste. Basidiospores 11.5–15 (−16) × 8–9.5 (−10) μm, heterodiametrical, angular 6–8 sided, somewhat nodulose, pinkish brownish, thick-walled.

In summer to autumn, scattered in rich humus on ground in mixed forests with *Pinus* and *Quercus*. Distributed in Northeast China.

1092 *Entoloma conferendum* (Britzelm.) Noordel

Pileus 2.3–5 cm broad, conical to hemispherical when young, conical-campanulate to conical-convex at maturity, slightly plane, with or without an inconspicuous papilla at centre, translucently striate reaching disc, light brown, beige brown to greyish brown, slightly darker at centre, glabrous, dry. Context greyish white, thin. Lamellae adnate, close, white at first, then pink. Stipe 3–7 cm long, 0.3–0.5 cm thick, cylindrical, gradually thicker downward, concolourous with pileus or lighter, light brown, with white pruinescence, longitudinally striate, with silky lustre, solid at first, then hollow, with white mycelium at base. Basidiospores 8.5–11.5 × 7.5–11 μm, various in shape, cuboid, rhomboic to stellate, thick walled, pinkish.

Gregarious on fallen wood or ground in broad-leaved forests. Distributed in Northeast and Northwest China.

1093 *Entoloma crassicystidiatum* T.H. Li & Xiao Lan He

Pileus 1–2.5 cm broad, conical-convex, subconical to subcampanulate when young, subplane when maturity, lacerated at margin, greyish brown, light brown to brown, dry, hygrophanous, closely with suberect flocculose fibrils or squamulose tomentum; subglabrous at margin at maturity, inconspicuously grooved. Context 1.5 mm thick, concolourous with pileus, with an indistinct odour. Lamellae adnate or sinuate, with a short decurrent tooth, lighter than pileus, greyish brown with pink, distant, thick, with 2 tiers of lamellulae. Stipe 2.5–3.5 cm long, 2–3 mm thick, cylindrical, light greyish brown, slightly lighter than pileus, longitudinally striate and densely tomentose. Basidiospores 11.5–14.5 × 7.5–9 µm, heterodiametric, 6–8 sided, obviously angular, thick walled, pinkish.

Scattered on sandy soil in mixed forests. Distributed in Southern China.

1094 *Entoloma crepidotoides* W.Q. Deng & T.H. Li

Pileus 5–15 mm broad, flabelliform to conchate, white when young, pinkish at maturity, with white tomentulum. Lamellae white at first, then pinkish, radiating from an attachment base, unequal, short decurrent. Stipe absent and only with a lateral attachment base. Basidiospores 8–9 × 6–7 µm, heterodiametric, 6–7 sided, pinkish. Pileipellis hyphae 3–9 µm diam., trichodermoid. Clamp connection absent.

Gregarious on ground in mixed forests. Distributed in Southern China.

1095 *Entoloma crocotillum* Xiao Lan He

Pileus 5–15 mm broad, campanulate, wide convex to plano-convex, fasciculately tomentose or fibrillose, white to cream at first, then orange white, yellowish white to light pink. Context thin, white. Lamellae adnate, with a short decurrent tooth, distant, white to pink. Stipe 4–7 cm long, 0.5–1.5 mm thick, cylindrical, hollow, white, translucent, glabrous, brittle. Basidia 35–48 × 11–14 µm, wide clavate, 4-spored. Basidiospores 9.5–12.5 × 7.5–9.5 µm, heterodiametric, 5–7 sided. Cheilocystidia 35–65 × 7–17 µm, cylindrical, subfusiform or wide clavate, thin-walled, hyaline. Clamp connection present.

On ground in mixed or bamboo forests. Distributed in Central China and Qinghai-Tibet Region, China.

1096 *Entoloma dysthaloides* Noordel.

Pileus 5–12 mm broad, convex, campanulate to conical-convex, sometimes plano-convex, translucently striate reaching to 3/4 length from margin to centre, greyish brown to light brown, with fibrillose squamules. Context thin, concolourous with pileus, with an indistinct odour. Lamellae adnate or sinuate, with a short decurrent tooth, greyish brown to light brown, distant to close, slightly serrate at edge. Stipe 1.5–6 cm long, 0.5–1.2 mm thick, cylindrical, concolourous with pileus or lighter, solid to hollow, greyish white or greyish brown fibrillose, with yellow brown bristles at base. Basidiospores 10.5–13 × 7–8.5 µm, heterodiametric, 5–7 sided, obviously angular, light pink to light brown.

Scattered on ground in broad-leaved forests (with *Acer*, *Betula*, *Quercus* and *Tilia*). Distributed in Northeast China.

Scattered on ground with mosses and fallen leaves at edge of coniferous forests. Distributed in Northeast China and Qinghai-Tibet Region, China.

1097 *Entoloma formosum* (Fr.) Noordel.

Pileus 2–5 cm broad, hemispherical, convex, campanulate to bamboo-hat shaped when young, truncated to inconspicuously depressed at centre, dark yellow, ferruginous yellow to yellowish brown, often with reddish tinge, minutely squamulose and radiately striate, less squamulose at margin, occasionally lacerated. Context concolourous with pileus or lighter, about 1 mm thick at stipe, with an inconspicuous odour and taste. Lamellae adnate, close, up to 6 mm broad, white at first, then pink, with 3–4 tiers of lamellulae. Stipe 5–9 cm long, 1.5–4.5 mm thick, cylindrical, central, hollow, lighter than pileus, glabrous, with white mycelium at base. Basidiospores 9.5–12.5 × 7–8.5 μm, heterodiametric, 5–6 sided, pinkish.

1098 *Entoloma furfuraceum* T.H. Li & Xiao Lan He

Pileus 4–10 mm broad, subhemispherical, obtusely convex to campanulate, then plano-prominent to wide convex, dry, brown, dark greyish brown or purplish brown, densely with fibrillose or furfuraceous squamules or suberect strigose hairs, which are dark grey, greyish brown with purplish tinge or concolourous with pileus, inconspicuously sulcate or striate. Context thin, concolourous with pileus. Lamellae up to 2.5 mm broad, adnate, with a short decurrent tooth, distant, thick, greyish brown, with 2 tiers of lamellulae. Stipe 3–5 cm long, 0.5–1 mm thick, cylindrical, concolourous with pileus or lighter, longitudinally striate, densely with grey to greyish brown furfuraceous squamules; base purplish brown, with bristles. Basidiospores 9–11 × 6.5–8 μm, 5–7 sided, heterodiametric, nodulose-angular, thick-walled, pinkish to light pinkish brown.

Scattered on ground with mosses in broad-leaved forests. Distributed in Northeast China.

1099 *Entoloma hainanense* T.H. Li & Xiao Lan He

Pileus 5–6 cm broad, convex, brown with a slight purplish tinge, finely tomentulose, irregularly wrinkled, not viscid, estriate or inconspicuously striate at margin. Context 4–4.5 mm thick at stipe, white, with an indistinct odour. Lamellae up to 5 mm broad, slightly thick, adnate or subsinuate, close, light yellow, pink at maturity, unequal, with 2 tiers of lamellulae. Stipe 8–9.5 cm long, 7–10 mm thick, 16–18 mm thick at base, subclavate, solid, white, with white mycelium at base. Basidiospores 9.5–11 × 8.5–10 μm, 5–6 sided, obviously angular, isodiametric to heterodiametric, thick-walled, light pink.

Scattered on ground in broad-leaved forests. Distributed in Southern China.

1100 *Entoloma incanum* (Fr.) Hesler

Pileus 1–1.5 cm broad, convex or subcampanulate, umbilicate at centre, yellowish green with greyish brown tinge, greenish brown to light yellowish brown with green tinge, radially striate reaching to centre, glabrous or with fine squamules. Context thin, white. Lamellae adnate, distant to close, white at first, pink or dirty pink at maturity. Stipe 3–6 cm long, 2–3 mm thick, cylindrical, hollow, bluish yellow, changing to bluish green when injured, with white mycelium at base. Basidiospores 11–14 × 8–10 μm, 6–8 sided, light pink.

Scattered or gregarious on ground in broad-leaved forests. Distributed in Northeast China and Qinghai-Tibet Region, China.

1101 *Entoloma kujuense* (Hongo) Hongo W & Izawa

Pileus 2.5–8 cm broad, convex or hemispherical when young, subplane at maturity, sometimes lacerated at margin, dark purple or purplish blue, with velvet tomentose or furfuraceous squamules, not viscid, estriate at margin. Context white, 3 mm thick at disc. Lamellae up to 5 mm broad, adnate, with a short decurrent tooth, white at first, then pink, thin or thick, with 2–3 tiers of lamellulae. Stipe 3–6 cm long, 0.3–0.7 mm thick, cylindrical, concolourous with pileus, densely with furfuraceous squamules, solid, with white mycelium at base. Basidiospores 10–12.5 × 7–8.5 μm, 6–7 sided, heterodiametric, sometimes nodulose-angular, light pink.

On ground in broad-leaved forests. Distributed in Central and Southern China.

1102 *Entoloma laccaroides* T.H. Li & Xiao Lan He

Pileus 1.5–4 cm broad, convex, umbilicate or subinfundibuliform at centre, yellowish white to pinkish flesh-coloured, conspicuously striate or shallowly grooved, serrate or undulate at margin. Lamellae adnate to short decurrent, 7 mm broad, thick, concolourous with pileus, then pink, changing to reddish brown when injured, irregular or subundulate at edge. Stipe 5–8 cm long, 3–6 mm thick, cylindrical, central, hollow, longitudinally striate, with white mycelium at base. Basidiospores 7.5–10.5 µm wide, cuboid, thick-walled, pink. Cheilocystidia 27–95 × 5.5–12 µm, cylindrical to subclavate, with light brown intracellular pigment, thin-walled. Pileipellis hyphae 8–25 µm diam, cylindrical, repent, with light yellow granular pigment. Clamp connection present.

On ground in broad-leaved forests. Distributed in Southern China.

1103 *Entoloma liaoningense* Yu Li et al.

Pileus 1–1.5 cm broad, hemispherical or convex to plano-umbonate, slightly depressed around umbo, often with a small umbo at central depressed portion, glabrous, radially slightly grooved from margin to disc, dirty yellow to light greyish brown, lighter at margin, darker at grooves. Context thin, greyish white. Lamellae sinuate or adnate with a decurrent tooth, distant, greyish white when young, pink at maturity, unequal. Stipe 3.5–4.5 cm long, 1–2.5 mm thick, cylindrical, hollow, light brown, darker than pileus, glabrous, with white mycelium at base. Basidiospores 8–10.5 × 7–8.5 µm, isodiametric, 5–6 sided, pinkish. Basidia often with two small sterigmata.

On ground in mixed or coniferous forests. Distributed in Northeast China.

This species is similar to *E. bisporum* in Japan, however the basidiospore of the latter species is heterodiametric.

1104 *Entoloma luteum* Peck

Pileus 10–25 mm broad, hemispherical, convex or subcampanulate, glabrous to fibrillose, with squamules at apex, without an umbo or acute apex, light yellow to dark yellow, changing to lighter with pinkish tinge at maturity, dry, hygrophanous, striate. Context white or light yellow, thin. Lamellae adnate or subfree, distant, white at first, then pink. Stipe 4.5–6.5 cm long, 3–5 mm thick, cylindrical, hollow, longitudinally striate, brittle. Basidiospores 7.5–9.5 µm, cuboid, pinkish.

Solitary or scattered on ground in forests. Distributed in Southern China.

1105 *Entoloma mastoideum* T.H. Li & Xiao Lan He

Pileus 2.5–7 cm broad, conical when young, then bamboo-hat shaped or subplane, cuspidal or papillate at centre, translucently striate or nearly estriate, flesh, pink or fleshy pink, lighter at margin, glabrous, often obviously overhanging lamellae at margin. Context thin, white. Lamellae up to 5 mm broad, adnate, close, white at first, then pink, with 3–4 tiers of lamellulae. Stipe 3.5–9 cm long, 3–8 mm thick, cylindrical, lighter than pileus, white, ivory to slightly yellow, translucent, hollow, brittle. Basidiospores 9.5–12 × 7–8 µm, heterodiametric, 6–8 sided, distinctly angular, thick-walled, pinkish.

Scattered to fasciculate on sandy soil in broad-leaved forests. Distributed in Southern China.

1106 *Entoloma mediterraneense* Noordel. & Hauskn.

Pileus 1.5–3.5 cm broad, convex, slightly depressed at centre, estriate or inconspicuously striate, dark grey to greyish brown, slightly greyish blue, nearly darkish brown at centre, with greyish brown squamules, with gradually less short fibrillose squamules at margin. Context 0.5 mm thick at centre, greyish white. Lamellae sinuate, with a short decurrent tooth, close, thin, white when young, pink at maturity, with 3 tiers of lamellulae, irregular at edge, concolourous with lamella side face or light blue. Stipe 45–55 mm long, 2.5–4 mm thick, cylindrical, hollow, nearly dirty white to dark greyish blue, short tomentose or pruinose, with white mycelium at base. Basidiospores 8–10.5 × 6–7.5 µm, heterodiametric, 5–6 sided, light pink.

On ground in broad-leaved forests. Distributed in Northeast China.

1107 *Entoloma murrayi* (Berk. & M.A. Curtis) Sacc.

Pileus 2–4 cm broad, bamboo-hat shaped to conical, with an prominently long acute apex or papilla at centre, glabrous or fibrillose, with a weak silky sheen at maturity, striate or shallowly sulcate, light yellow to yellow or bright yellow, sometimes with lemon yellow tinge. Context thin, nearly translucent. Lamellae up to 5 mm broad, adnate or sinuate, distant, with 2–3 tiers of lamellulae, concolourous with pileus to pinkish. Stipe 4–8 cm long, 2–4 mm thick, cylindrical, glabrous to fibrillose, yellowish white, light yellow to concolorous with pileus, striate, hollow, slightly enlarged downward. Basidiospores 7–9.5 μm, cuboid, thick-walled, light pink.

In summer to autumn, solitary to gregarious on ground in mixed forests. Distributed in Central and Southern China.

1108 *Entoloma omiense* (Hongo) E. Horak

Pileus 3–4 cm broad, conical at first, then bamboo-hat shaped to subcampanulate, slightly acute or subobtuse at centre, light greyish brown to light yellowish brown, striate, glabrous. Context thin, white. Lamellae 5–7 mm broad, adnate, close, thin, white when young, pink to light pinkish yellow at maturity, with 2–3 tiers of lamellulae. Stipe 5–14 cm long, 3–4 mm thick, cylindrical, whitish to concolourous with pileus, glabrous, with white mycelium at base. Basidiospores 9.5–12.5 × 9–11.5 μm, 5–6 sided, normally 5 sided, isodiametric to subisodiametric, pinkish.

Solitary or scattered on ground. Poisonous. Distributed in Southern and Central China.

1109 *Entoloma pallidocarpum* Noordel. & O.V. Morozova

Pileus 8–13 cm broad, convex to nearly plane, hygrophanous, translucently striate at margin, beige yellow to light yellow brown, glabrous. Context white. Lamellae adnate, close, white when young, then pink. Stipe 5–8 cm long, 13–17 mm thick, cylindrical, white, solid at first, then hollow, with white mycelium at base. Basidiospores 7.5–9 × 6.5–8 μm, 6–7 sided, isodiametric to subisodiametric, thick-walled, pinkish.

On ground in mixed forests. Distributed in Northeast China.

1110 *Entoloma parvifructum* Xiao Lan He et al.

Pileus 5–15 mm broad, conical to subcampanulate, conspicuously umbonate at centre, conspicuously striate or subsulcate, light brown to greyish orange brown, dark greyish brown at striae or sulci. Context thin, with inconspicuous odour and taste. Lamellae adnate, with a decurrent tooth, up to 1.5 mm broad, distant, grey at first, pink at maturity, sterile at edge. Stipe 5–6 cm long, 0.5–1.5 cm thick, cylindrical, central, hollow, light grey, glabrous, with white mycelium at base. Basidia 37–48 × 18–25 μm, broadly clavate. Basidiospores 10–14.5 μm, cuboid, thick-walled, pinkish. Cheilocystidia 40–65 × 8–17 μm, hyaline, thin-walled. Pileipellis hyphae repent, terminal elements 50–200 × 6–15 μm, cylindrical or subfusiform. Clamp connection present.

On ground in broad-leaved forests. Distributed in Southern and Central China.

1111 *Entoloma petchii* E. Horak

Pileus 1.5–3.5 cm broad, subhemispherical to convex at first, plane at maturity, dark greyish brown to greyish black, with blackish brown squamules, estriate at margin. Context greyish brown, thin. Lamellae up to 5 mm broad, adnate, subclose, slightly thick, brown with greyish tinge at first, pinkish brown to greyish brown with pinkish tinge, darkish brown at edge, with 2 tiers of lamellulae. Stipe 2–4 cm long, 2–3 mm thick, cylindrical, with blackish brown squamules, hollow, with white mycelium at base. Basidiospores 9.5–12 μm broad, cuboid, thick-walled, pinkish.

In autumn, solitary on ground in broad-leaved forests. Distributed in Southern China.

1112 *Entoloma praegracile* Xiao Lan He & T.H. Li

Pileus 0.8–2 cm broad, convex at first, then plane, slightly depressed or plane at centre, light yellow, light yellow tinged with pink or orange yellow, with conspicuous orange red tinge when dried, hygrophanous, translucently striate reaching to disc, glabrous. Context thin, concolourous with pileus. Lamellae up to 1 mm broad, adnate with a short decurrent tooth, distant, white at first, then changing to pink, with 1–2 tiers of lamellulae. Stipe 4–5 cm long, 1–1.5 mm thick, cylindrical, concolourous with pileus or darker, orange yellow, glabrous, hollow, brittle, with white mycelium at base. Basidiospores 9–10.5 × 6.5–8 μm, 5–6 sided, heterodiametric, sometimes weakly angular, thick-walled, light pink.

Scattered to fasciculate on ground in broad-leaved forests. Distributed in Central and Southern China.

1113 *Entoloma quadratum* (Berk. & M.A. Curtis) E. Horak

Pileus 1–6 cm broad, conical to subcampanulate, sometimes with a conspicuously acute apex, orange yellow, orange red, salmon to orange brown, glabrous, striate or sulcate reaching to centre. Context up to 1 mm thick at stipe, concolourous with pileus. Lamellae up to 3 mm broad, sinuate or adnate, distant, concolourous with pileus, slightly undulate at edge, with 2 tiers of lamellulae. Stipe 3–6 cm long, 2–4 mm thick, cylindrical, hollow, fibrous to cartilaginous, longitudinally striate, mostly concolourous with pileus. Basidiospores 7.5–10.5 μm, cuboid, light pink.

Solitary to scattered on ground in broad-leaved forests. Distributed in Southern and Central China.

1114 *Entoloma* cf. *sarcopum* Nagas. & Hongo

Pileus 6–9 cm broad, conical at first, then hemispherical to subplane, plane and prominent at centre, smooth, mouse-greyish brown, with white mercerized striae, then as mouse-greyish brown striae, incurved at margin. Context thin, white. Lamellae sinuate to subfree, broad, unequal, dirty white to pink, subundulate or serrate at edge. Stipe 9–14 cm long, 0.9–1.2 cm thick, equal or thicker downwards, white, even, solid. Basidiospores 9.5–12.5 × 7–8.5 μm, wide elliptically angular, smooth, pinkish.

In summer to autumn, gregarious or solitary on ground in broad-leaved and mixed forests. Edible. Distributed in Northeast and Northwest China.

1115 *Entoloma sepium* (Noulet & Dass.) Richon & Roze

Pileus 3–10 cm broad, convex to plano-prominent to nearly plane, tomentose or finely fibrillose when young, subglabrous at maturity, grey, greyish brown to light yellowish brown, changing to light orange red to ferruginous when injured. Context white. Lamellae adnate, close, white to pink, changing to orange red to ferruginous when injured. Stipe 6–10 cm long, 0.7–1.5 cm thick, thicker at base, greyish white to greyish brown, changing to orange yellow to orange red when injured, glabrous or with fibrillose, with white mycelium at base. Basidiospores 8.5–11 × 8–10.5 μm, isodiametric, 5–6 sided, pinkish.

On ground of forests. Edible. Distributed in Northeast and Northern China.

This species is similar to *E. clypeatum*, however the latter species is unchanging when injured.

1116 *Entoloma shandongense* T. Bau & J.R. Wang

Pileus 1–1.5 cm broad, conical or hemispherical, slightly depressed at centre, non-hygrophanous, glabrous, blue slightly with purple tinge, sparsely tomentose, translucently striate and lighter at margin. Context thin, white to light brown. Lamellae subdecurrent, distant, light beige yellow to pinkish. Stipe 1–1.5 cm long, 1–2 mm thick, central to slightly eccentric, cylindrical to subclavate, gradually thinner at base, concolourous with lamellae, light beige yellow to light pink, light brown to white at base. Basidiospores 5.6–7.2 × 8.3–10 µm, elliptically angular, 5–8 sided, thick-walled, pinkish.

On ground in broad-leaved forests. Distributed in Northern China.

1117 *Entoloma strictius* (Peck) Sacc.

Pileus 2–6 cm broad, conical, sometimes subcampanulate or slightly expanded, with a small papilla at centre, greyish white, greyish brown or light greyish brown, glabrous, inconspicuously to slightly conspicuously striate, even at margin. Context thin, concolourous with pileus. Lamellae adnate, white to pink, close, undulate at edge. Stipe 6.5–20 cm long, 3–6 mm thick, cylindrical, hollow, longitudinally striate, twisted, with white mycelium at base. Basidiospores 10–13 × 7.5–11 µm, 5–6 sided, heterodiametric, light pink.

Scattered or gregarious on ground in broad-leaved forests. Distributed in Central China.

1118 *Entoloma stylophorum* (Berk. & Broome) Sacc.

Pileus 0.7–1.4 cm broad, conical or convex at first, then plano-prominent or plane, with a conspicuously acute or papillate apex, white, yellowish white or white slightly tinged with grey to with light pink, with mercerized fibrils. Context thin, white. Lamellae up to 2 mm broad, adnate, distant to close, thin, with 2 tiers of lamellulae, white at first, then changing to pink. Stipe 3–4 cm long, 1–2 mm thick, hollow, white or with creamy tinge, glabrous or tomentulose, with white mycelium at base. Basidiospores 9.5–11 × 7.5–9 µm, heterodiametric, 5–7 sided, pinkish.

Scattered on ground in broad-leaved forests. Distributed in Southern China.

1119 *Entoloma subaraneosum* Xiao Lan He & T.H. Li

Pileus 1.5–2 cm broad, subconvex to campanulate when young, truncated at apex, then slightly expanded, light brown to brown, densely with appressed argenteous white or mouse grey fibrils, estriate. Lamellae adnate or sinuate, subventricose, subdistant, greyish white at first, then grey or light brown with pink tinge. Stipe 4.5–6 cm long, 2–3 mm thick, cylindrical, concolourous with pileus or lighter, hollow, longitudinally striate, with greyish white or argenteous grey fibrils, with yellowish brown or yellowish grey mycelium or long fibrils at base. Basidiospores 9.5–12.5 × 7.5–8.5 µm, 5–7 sided, heterodiametric, obviously angular, thick-walled, pink to pinkish brown.

In summer to autumn, scattered on ground in mixed forests. Distributed in Northeast China.

1120 *Entoloma subclitocyboides* W.M. Zhang

Pileus 10–13 cm broad, clitocyboid or infundibuliform, with short pubescence at first, then glabrescent, beige yellow, dirty yellow to light yellowish brown, gradually lighter from centre to margin, with light brown fibrils or squamules, glabrous at margin, lacerated at margin at age. Context up to 2 mm thick, white. Lamellae up to 8 mm broad, adnate, very close, white at first, changing to pink at maturity, unequal. Stipe 6–7 cm long, 1.1–1.8 cm thick, beige or dirty white, lighter than pileus, striate, with minute brown granules at apex, with white mycelium at base. Basidiospores 7–9 × 7–8.5 µm, 4–5 sided, mostly 5 sided, isodiametric, pinkish.

Solitary or scattered on ground in broad-leaved forests. Distributed in Southern and Central China.

1121 *Entoloma subtenuicystidiatum* Xiao Lan He & T.H. Li

Pileus 0.8–2.5 cm broad, hemispherical to convex, striate or with sulcate, depressed and squamulose at centre, subglabrous elsewhere, yellow to dark beige, slightly with reddish brown tinge at centre. Context thin, submembranous. Lamellae up to 3 mm broad, adnate, close, white at first, then pink, pinkish brown or concolourous with pileus at edge, with 3 tiers of lamellulae. Stipe 4–6 cm long, 1–2 mm thick, cartilaginous, translucent, white or light brown with pinkish, glabrous, with white mycelium at base. Basidiospores 9.5–13 × 7.5–9 µm, heterodiametric, 6–9 sided, pinkish.

Scattered on grassland with *Zoysia tenuifolia* and *Cynodon dactylon*. Distributed in Central China.

1122 *Entoloma sulcatum* (T.J. Baroni & Lodge) Noordel. & Co-David

Pileus 7–15 mm broad, plane or plano-depressed, sparsely tomentose or fibrillose, cream or whitish at first, then changing to light pink or orangish white, slightly sulcate when young, estriate or inconspicuously sulcate at maturity. Context thin, white. Lamellae adnate or adnate with a short decurrent tooth, white to pink. Stipe 1.5–3 cm long, 0.5–2 mm thick, cylindrical, glabrous, concolourous with pileus, often translucent. Basidia 26–35 × 9–11 µm, clavate, 4-spores. Basidiospores 9.5–12.5 × 7.5–9.5 µm, heterodiametric, 5–6 sided. Cheilocystidia 35–60 × 9–20 µm, cylindrical to clavate, hyaline, thin-walled. Pleurocystidia absent. Clamp connection absent.

On ground in mixed forests. Distributed in Southern China.

1123 *Entoloma talisporum* Corner & E. Horak

Pileus 1–2.5 cm broad, convex to plano-depressed, dry, tomentose, radiately striate, even at margin, sometimes incurved, white, creamy white to white with yellow tinge, darker at centre. Context thin, white to yellow. Lamellae 8–14 per cm at pileal margin, adnate to subdecurrent, unequal, white to yellowish white, then pink. Stipe 2–6 cm long, 1–3 mm thick, cylindrical, with white mycelium and slightly enlarged at base, hollow, fleshy, white to yellowish white. Basidiospores 6–8 µm broad, cuboid, apiculate, smooth, pinkish.

Scattered or fasciculate on ground in mixed forests. Distributed in Southern China.

This species is similar to *E. indicum*, however the latter species has larger basidiospores.

1124 *Entoloma tenuissimum* T.H. Li & Xiao Lan He

Pileus 3–8 mm broad, obtusely conical, convex, broadly convex to subcampanulate, dry, conspicuously sulcate or translucently striate, greyish brown to light brown with pink tinge, changing to lighter at margin, darker at sulci, fibrillose. Context thin, whitish. Lamellae up to 1 mm broad, adnate to with a decurrent tooth, greyish white at first, light brown with pinkish at maturity, with 2 tiers of lamellulae. Stipe 2.5–4 cm long, 0.3–1 mm thick, concolourous with pileus or lighter, cartilaginous, sparsely short tomentose, with light yellowish brown and long bristles at base. Basidiospores 14–18.5 × 9–11.5 µm, heterodiametric, 6–9 sided, pinkish to light brown.

Scattered on ground in broad-leaved forests. Distributed in Northeast China.

1125 *Entoloma tubaeforme* T.H. Li et al.

Pileus 1.6–4 cm broad, conspicuously depressed at centre, infundibuliform to trumpet-shaped, radially fibrillose or striate, light grey orange brown to dark brown, gradually lighter toward margin, with dark brown squamules at central depression, gradually glabrescent toward margin, dry. Context white, thin. Lamellae up to 6 mm broad, decurrent, white at first, pink at maturity, close, unequal, with lamellulae. Stipe 2.2–4 cm long, 2–4 mm thick, cylindrical, central, whitish to concolourous with pileus, sometimes hygrophanous, white tomentose, with white mycelium at base. Basidiospores 8–11.5 × 6.5–9 μm, heterodiametric, 4–6 sided, thick-walled. Cheilocystidia 55–162 × 10–22 μm, cylindrical to clavate or lageniform. Pleurocystidia abscent. Clamp connections present.

Gregarious on ground in forests of *Casuarina equisetifolia*. Distributed in Southern China.

1126 *Eotoloma virescens* (Berk. & M.A. Curtis) E. Horak

Pileus 2–3 cm broad, conical or convex when young, slightly expanded at maturity, fibrillose-squamulose to fibrillose, blue, bluish green, changing to green to greenish brown when injured, inconspicuously striate. Context 0.2–1 mm thick, concolourous with pileus. Lamellae 4–6 mm broad, sinuate, distant, blue or concolourous with pileus when young, slightly pinkish at maturity, with 2–3 tiers of lamellulae. Stipe 4–7 cm long, 3–5 mm thick, longitudinally striate or fibrillose, concolourous with pileus or lighter, changing to green to greenish brown when injured, slightly enlarged at base, with white basal mycelium. Basidiospores 9–12 μm, cuboid, conspicuously apiculate, pinkish.

In summer to autumn, solitary or gregarious on lawn in forests. Distributed in Central and Southern China.

Agarics 3: Species with Names Beginning with F-I

1127 *Favolaschia manipularis* (Berk.) Teng

Pileus 1–3 cm broad, hemispherical to oblate hemispherical, white in upper portion and translucent in lower portion when young, then discolouring to dirty white to brown gradually, with a translucently reticulate pattern corresponding to the underneath pore ridges. Context thin, concolourous with pileus. Tubes 1.5–5 mm deep, porous, adnate, lighter than pileus or dirty white, waxy; pores 0.5–1 mm wide, angular. Stipe central, 2–5 cm long, 1.5–3 mm thick, cylindrical, hollow, concolourous with pileus or lighter, brittle, pruinose to glabrous. Basidiospores 6–8 × 4–5 µm, oval to wide elliptical, smooth, hyaline, amyloid.

Gregarious on dead wood. Distributed in Southern China.

© Central China Farmer's Publishing House 2025

Y. Li et al., *Atlas of Chinese Macrofungal Resources*, https://doi.org/10.1007/978-981-96-0502-6_4

1128 *Favolaschia pustulosa* (Jungh.) Kuntze

Pileus up to 3.2 cm broad, 2 mm thick at base, flabelliform to semi-circular, projecting up to 2 cm from substrate, with a translucently reticulate pattern corresponding to underneath hymeniphore, yellowish white, cream to light yellow, fleshy when fresh, gelatinous when dried; margin acute, incurved when dried. Hymeniphore porous. Pores larger at base and becoming smaller toward margin, 0.5 per mm at base, 1–2 per mm at margin, angular, cream to light yellow when dried. Context 0.1 mm thick, cream to brown, gelatinous. Basidiospores 6.5–8 × 6–7 μm, globose to subglobose, thick walled, hyaline, amyloid, cyanophilic.

In summer to autumn, on rotten woods of broad-leaved trees, causing white rot in wood. Distributed in Southern China.

1129 *Favolaschia tonkinensis* (Pat.) Kuntze

Pileus 1–2 cm broad, flabelliform, white to beige, uneven, with reticulate ornamentation corresponding to underneath hymeniphore, viscid. Hymeniphore tubular, white; pores subcircular. Stipe absent or very short. Basidia 30–40 × 6–8 μm. Basidiospores 8–12 × 7–10 μm, wide elliptical to subglobose, smooth, hyaline, amyloid.

In summer to autumn, on rotten stems of bamboo. Distributed in Southern China.

1130 *Flammulaster erinaceellus* (Peck) Watling

Pileus 1–4 cm broad, obtusely conical to convex at first, then nearly plane, rarely umbonate, dark ferruginous brown, densely with granular to pulvinate khaki fagacious squamules, appendiculate with fagacious velar remnants at margin. Context thin, olive yellow to dark yellow. Lamellae adnate, greyish white to light cinnamon brown, changing to light ochraceous brown, then dark yellowish brown, close. Stipe 3–5 cm long, 3–4 mm thick, cylindrical, greyish white and white powdery above annulus, with squamules similar to those on pileus below annulus. Annulus fibrillose, rarely submembranous. Basidiospores 6.5–8 × 4–5 μm, elliptical, smooth, light brown.

In spring to autumn, scattered or gregarious on rotten woods of broad-leaved trees. Distributed in Northeast China.

1131 *Flammulina rossica* Redhead & R.H. Petersen

Pileus 1.5–5 cm broad, flat to plane, white, beige to light yellow, darker at disc, viscid when wet. Context thin, white. Lamellae sinuate, white to beige. Stipe 1.5–4 cm long, 2–5 mm thick, yellowish brown at apex, dark brown to nearly black at lower portion, tomentose, not viscid. Basidiospores 7.5–11 × 4–4.5 μm, elliptical to long elliptical, smooth, hyaline to light yellow, inamyloid.

In summer to autumn, on rotten woods of *Salix cupularis*. Edible. Distributed in Qinghai-Tibet Region, China.

1132 *Flammulina velutipes* (Curtis) Singer

Pileus 1.5–7 cm broad, oblate hemispherical when young, then flat to plane, light yellowish brown to yellowish brown, darker at disc, creamy yellow and striate at margin, slightly viscid when wet. Context thick at centre, thin at margin, white, soft. Lamellae sinuate, white to beige, close, unequal. Stipe 3–7 cm long, 0.2–1 cm thick, cylindrical, yellowish brown at apex, dark brown to nearly black at lower portion, tomentose, not viscid, fibrillose, soft insides, then hollow, with a rooting base as a pseudorhiza growing together with other stipe bases. Basidiospores 8–12 × 3.5–4.5 μm, elliptical to long elliptical, smooth, hyaline or light yellow, inamyloid.

In early summer to late autumn or early winter, fasciculate on rotten stumps and roots of broad-leaved trees, its pseudorhiza is usually from rotten woods buried in soil. Edible, famous cultivated edible mushroom, product name is needle mushroom or enoki mushroom, with white and brown varieties. Distributed widely in China.

1133 *Flammulina yunnanensis* Z.W. Ge & Zhu L. Yang

Pileus 1.5–3.5 cm broad, flat to plane, yellow to creamy yellow, melline yellow to light orange at centre. Context thin, whitish to pileus colour. Lamellae sinuate, white to beige. Stipe 3–6 cm long, 3–7 mm thick, tomentose. Basidiospores 5.5–6.5 × 3–4 μm, elliptical, smooth, hyaline, amyloid.

In summer to autumn, on rotten woods in subtropical broad-leaved trees. Edible. Distributed in Central China.

1134 *Floccularia albolanaripes* (G.F. Atk.) Redhead

Pileus 3.5–6.5 cm broad, flat to plane, yellow to bright yellow, with light brown squamules, darker and obviously umbonate at disc. Context white, unchanging when injured. Lamellae sinuate, beige to light yellow. Stipe 5–8 cm long, 0.5–1 cm thick, white and glabrous at apex, beige to light yellow at central and lower portions, with yellow flocculose recurved scales. Basidiospores 6–7.5 × 4–5 μm, elliptical, smooth, hyaline, weakly amyloid.

In summer to autumn, on ground in forests. Distributed widely in China.

1135 *Floccularia luteovirens* (Alb. & Schwein.) Pouzar

≡ *Armillaria luteovirens* (Alb. & Schwein.) Sacc.

Pileus 5–13 cm broad, oblate hemispherical, convex to plane, slightly prominent to non-prominent at centre, sulfureous yellow, yellow to bright yellow, easily discolouring to whitish in dry conditions, with flocculent scales or cracks, or with yellowish green to yellow recurved coarse scales, incurved at margin. Context thick, white. Lamellae close, adnate or sinuate, beige to yellow, or similar to pileus, often yellow, unequal. Stipe 3.5–9 cm long, 1–2.6 cm thick, white to with yellow, solid, with yellow scales below annulus. Annulus superior, yellow. Basidiospores 6–7 × 4–5 μm, elliptical, nearly smooth to very inconspicuously asperulate, hyaline, weakly amyloid.

In summer to autumn, on prairie or alpine grassland. Edible. Distributed in Northern, Northwest and Qinghai-Tibet Region, China.

1136 *Galerina autumnalis* (Peck) A.H. Sm. & Singer

Pileus 1.5–4.5 cm broad, hemispherical, campanulate to plane, often papillate or umbonate at centre, ochraceous, yellowish brown to brown, viscid when wet, translucently striate, hygrophanous. Context thin, brown. Lamellae up to 5 mm wide, adnate or decurrent, subclose, concolourous with pileus at first, then becoming darker to ferruginous. Stipe 5.5–8 cm long, 3–5 mm thick, cylindrical, hollow, ferruginous, slightly lighter in upper portion than in lower portion, with a fibrillose and fagacious annulus at upper portion, with white mycelium at base. Basidiospores 8–9.5 × 5–6 μm, elliptical, inequilateral, smooth, inamyloid, without a germ-pore, inconspicuously rugulose or finely asperulate, with an obvious smooth supraphilar depression, brown to dark brown.

In late autumn, gregarious on rotten woods. Poisonous. Distributed in Northeast, Northwest and Qinghai-Tibet Region, China.

1137 *Galerina fasciculata* Hongo

Pileus 2–5 cm broad, hemispherical at first, then plane, not viscid, glabrous, light cinnamon, hygrophanous, becoming light yellow from centre to margin when being dried. Context thin, white to light yellow. Lamellae adnate or decurrent, cinnamon, slightly pruinose at edge, close or distant. Stipe 6–9 cm long, 2.5–5 mm thick, light yellow to light clay, hollow, fibrillose, pruinose at apex, with white mycelium at base. Annulus medial and inferior, dirty brown, membranous, fibrillose, fagacious and then vanished. Basidiospores 6–9 × 4–5 μm, long elliptical, elliptical to oval, verruculose except smooth supraphilar depression, without an apical germ-pore, reddish brown to brown, inamyloid.

In autumn, solitary to clustered on ground in forests. Poisonous. Distributed in Southern China.

This species is similar to *G. autumnalis*. However, the annulus of this species is membranous and medial-inferior, while the annulus of the latter species is fibrillose and superior.

1138 *Galerina helvoliceps* (Berk. & M.A. Curtis) Singer

Pileus 1–4 cm broad, hemispherical to plane, sometimes papillate or umbonate at centre, glabrous, beige yellow, yellow to ochraceous yellow, hygrophanously striate when wet and incurved at margin, not viscid. Context thin, white or dirty white. Lamellae 1–4 mm wide, adnate, decurrent or sinuate, unequal, forked, subdistant, dirty yellow, ochraceous yellow or yellowish brown. Stipe 1.5–7 cm long, 1–7 mm thick, straight or curved, hollow, sometimes lighter to light yellow at upper portion, dark brown at lower portion, with white tomentose mycelium at base. Annulus superior, dirty white to yellow, membranous, thin. Basidiospores 8–11 × 5–6.5 μm, elliptical to amygdaliform, verruculose or rugulose, with an obvious smooth supraphilar depression, without a germ-pore, ferruginous brown, inamyloid.

Gregarious or scattered on rotten woods in mixed forests. Poisonous. Distributed in Northeast China.

This species is similar to *G. marginata*, but former has a narrow and incomplete annulus, while latter has not. Former has slightly larger basidiospores than latter.

1139 *Galerina sulciceps* (Berk.) Boedijn

Pileus 1–3 cm broad, flat to plane, yellow brown, slightly depressed and papillate at centre, undulate at margin, obviously radiately sulcate reaching to centre. Context thin, whitish to light brown. Lamellae sinuate, light brown, distant. Stipe 3–5 cm long, 3–5 mm thick, yellow at apex, becoming darker downwards, blackish brown at base. Annulus absent. Basidiospores 7.5–10 × 4.5–5 μm, elliptical to amygdaliform, verruculose, with a calyptrate perispore, ferruginous brown.

In summer to autumn, on humus or rotten woods in tropical to south subtropical forests. Highly poisonous. Distributed in Southern and Central China.

1140 *Galerina vittiformis* (Fr.) Earle

Pileus 0.8–1.5 cm broad, conical, campanulate or plane, sometimes acutely umbonate at disc, yellowish brown, glabrous, radially striate from centre to margin, inconspicuously striate when dried. Context thin. Lamellae adnate, distant, yellowish brown. Stipe 2.5–3 cm long, 1–1.5 mm thick, cylindrical, reddish brown, with fine fibrils concolourous with pileus at upper portion, dark reddish brown at lower portion, hollow. Basidiospores 9–12 × 5.5–7 μm, long elliptical, minutely verruculose or aperulate, smooth on supraphilar depression, without a germ-pore, ferruginous brown, inamyloid.

Scattered on mosses, or rotten woods covered with moss in mixed forests. Poisonous. Distributed in Northeast, Inner Mongolia and Northwest China.

This species is similar to *G. hypnorum*, but the pileus of the latter species is only hygrophanously striate at edge, and has an undeveloped veil, and its basidiospores are elliptical or oval to amygdaliform.

1141 *Gymnopilus aeruginosus* (Peck) Singer

Pileus 2.5–7 cm broad, plane, dirty white to vinaceous, partially light green, with dark brown scales. Context light yellow to beige, bitter. Lamellae brownish yellow to light ferruginous brown. Stipe 2–3 cm long, 2–5 mm thick, subcylindrical, brown to purplish brown, with small fibrillose squamules, with beige to light yellow mycelium at base. Annulus membranous, light brown. Basidiospores 7–8.5 × 4–5 μm, elliptical to oval, verruculose, without a germ-pore, ferruginous brown.

In summer to autumn, on rotting wood in forests. Poisonous. Distributed widely in China.

1142 *Gymnopilus aurantiobrunneus* Z.S. Bi

Pileus 1.8–5 cm broad, oblate hemispherical to plane, light yellow to yellowish brown or ferruginous brown to purplish brown, not viscid, tomentose and squamulose. Context 1–5 mm thick at stipe, white or yellow to pinkish yellow or light yellow, unchanging when injured. Lamellae yellowish brown or ferruginous brown, unequal. Stipe 0.1–3.7 cm long, 1–6 mm thick, central to eccentric, squamulose or fibrillose, yellowish brown or purplish brown. Basidiospores 5–8 × 4–5 μm, elliptical, verruculose to nearly smooth, without a germ-pore, ferruginous brown.

In summer to autumn, scattered or gregarious on rotten wood in broad-leaved forests. Distributed in Southern China.

1143 *Gymnopilus dilepis* (Berk. & Broome) Singer

Pileus 3–7 cm broad, plane, purplish brown, with brown to dark brown erect scales at center. Context light yellow to beige, bitter. Lamellae brownish yellow to light ferruginous brown. Stipe 4–7 cm long, 0.3–1 mm thick, subcylindrical, brown to purplish brown, with small fibrillose scales. Annulus cortinoid, fagacious. Basidiospores 6–8.5 × 4.5–6 µm, elliptical to oval, verruculose, without germ-pore, ferruginous brown.

In summer to autumn, on rotten wood in forests. Poisonous. Distributed in Southern China.

1144 *Gymnopilus liquiritiae* (Pers.) P. Karst.

Pileus 3.5–5 cm broad, hemispherical to subcampanulate at first, plane in later stage, depressed at centre, moist, glabrous, light yellow, maize yellow to orange yellow, striate at margin. Context thin, bitter, yellow. Lamellae yellow or yellowish ferruginous at first, then changing to cinnamon, narrow, close, subadnate, unequal. Stipe 4.5–7 cm long, 4–5 mm thick, cylindrical, slender or tapering upwards, slightly curved, concolourous with pileus, light yellow or nearly dirty white, hollow, longitudinally fibrillose-striate, slightly enlarged at base, with white tomentulose basal mycelium. Annulus membranous or cortinoid, fagacious. Basidiospores 7–8.5 × 4–5.5 µm, amygdaliform or elliptical, rough, finely verruculose or asperulate, light yellow or light ferruginous.

In summer to autumn, gregarious or subfasciculate on rotten wood and dead wood in coniferous forests. Poisonous. Distributed in Northeast, Northwest, and Qinghai-Tibet Region, China.

1145 *Gymmopilus spectabilis* (Fr.) Singer

Pileus 3–8 cm broad, plane to flat, orange yellow to orange red, slightly darker at disc, with brown to light brown fibrillose scales, scales fagacious or easily washed away by rain. Context 10 cm thick, yellow to light yellow, bitter. Lamellae sinuate to subadnate, close, yellow, yellowish brown to ferruginous brown. Stipe 4–8 cm long, 5–10 mm thick, subcylindrical, solid; light yellow to yellow, with brown to light brown fibrillose scales, often tapering at base. Annulus superior, membranous, yellow to yellowish brown, ferruginous brown for numerous basidiospores deposited on upper surface. Basidiospores 7–9.5 × 5–6.5 µm, elliptical, ferruginous brown, verruculose. Cystidia vase-shaped.

In summer to autumn, on rotten wood in forests. Poisonous. Distributed in Northeast, Northwest, Central, and Qinghai-Tibet Region, China.

1146 *Gymnopus androsaceus* (L.) J.L. Mata & R.H. Petersen

≡ *Marasmius androsaceus* (L.) Fr.

Pileus 0.5–1.5 broad, hemispherical, convex to plane, slightly depressed at centre to umbilicate, radially sulcate, light brown, yellow brown, grey brown to dark brown, glabrous. Context thin, creamy. Lamellae adnate, crowded to subdistant, unequal, narrow, dull white to cinnamon, darker with age. Stipe 2.5–6.5 cm long, 0.3–1 mm thick, glabrous, dark brown or paler, pale reddish brown at upper part, nearly black at lower part, sometimes with pale yellow basal tomentose mycelium. Rhizomorphs usually present, 0.5–1 mm thick, slender, dark brown to black. Basidiospores 5–8.5 × 3–4.5 μm, long ellipsoid, hyaline, inamyloid.

In early summer to autumn, on debris of plants, especially on dead twigs in dark and humid environments, often abundantly occurring after raining. Medicinal. Distributed in the Northeast China, Central China and South China.

1147 *Gymnopus aquosus* (Bull.) Antonín & Noordel.

Pileus 0.4–4 cm in diam., convex when young, gradually expanded with age, golden yellow, darker at disc, glabrous, hygrophanous at margin. Context pale yellow, unchanging when injured. Lamellae adnate to subdecurrent, subclose, pale yellow, unequal, nearly even at edge. Stipe 3–11 cm long, 1–3 mm thick, cylindrical, central, slightly curved, concolorous with pileus. Basidiospores 5.5–7 × 3–4 μm, ellipsoid, smooth, pale yellow to hyaline, inamyloid.

In summer, caespitose in mixed forests with coniferous or broadleaf trees. Distributed in Northeast China.

1148 *Gymnopus confluens* (Pers.) Antonin et al.

≡ *Collybia confluens* (Pers.) P. Kumm

Pileus 1.5–4 cm in diam., campanulate to convex, then gradually expanded, slightly prominent at centre, glabrous to radially striate or fibrillose, light brown to pale reddish brown. Context thin, light brown. Lamellae sinuate to free, crowded, narrow, unequal, pale greyish brown to beige, white at edge. Stipe 4–8.5 cm long, 3–6 mm thick, central, cylindrical, smooth or grooved, pale reddish brown, darker towards base, with white basal mycelium. Basidiospores 5.7–8.6 × 3.1–4.4 μm, ellipsoid, smooth, hyaline, inamyloid.

In summer or autumn, gregarious or subcaespitose on decayed sticks and debris in forests. Distributed in most regions of China.

1149 *Gymnopus dryophilus* (Bull.) Murrilll

≡ *Collybia dryophila* (Bull.) P. Kumm.

Pileus 2–7 cm in diam., convex when young, then plane, ochraceous to pale brown, darker at disc, glabrous, event to subundulate at margin, hygrophanous. Context white, unchanging when injured. Lamellae free, subclose, dirty white to pale yellow, unequal, entire at edge. Stipe 3–7 cm long, 0.3–5 mm thick, cylindrical, fragile, tawny. Basidiospores 4.3–6.3 × 2.7–3.2 μm, ellipsoid, smooth, hyaline, inamyloid.

In summer to autumn, caespitose on humus ground in forests. Edible. Distributed in most regions of China.

1150 *Gymnopus menehune* Desjardin et al.

Pileus 0.8–2 cm in diam., convex when young, gradually expanded to slightly depressed at centre, glabrous, darker at centre, pale pinkish brown to pale brown, paler towards margin; margin slightly incurved when young, then expanded, slightly striate or rugulose. Context thin, concolorous with pileus to nearly white. Lamellae adnate to subdecurrent, subclose, nearly white or creamy white to tinted with the colour of pileus. Stipe 2.8–5.6 cm long, 1–2 mm thick, cylindrical, hollow, dry, nearly concolorous with pileus at apex, becoming darker to dark brown downwards. Basidiospores 6–8 × 3–5 μm, subellipsoid to pip-shaped, smooth, hyaline, inamyloid.

Caespitose on humus or fallen leaves of *Casuarina equisetifolia* or other broad leaved trees. Distributed in South China.

1151 *Gymnopus peronatus* (Bolton) Gray

≡ *Collybia peronata* (Bolton) P. Kumm.

Pileus 2.5–5 cm in diam., convex when young, then gradually applanate, often inflexed at margin, often slightly umbonate at centre, pale brown to tawny, darker at disc, gradually paler towards margin, eroded and subhygrophanous at margin. Context tawny, unchanging when injured. Lamellae free, subclose, tawny, unequal, entire at edge. Stipe 5–8 cm long, 3–5 mm thick, cylindrical, central, concolorous with pileus, minutely squamulose, with thick yellowish basal mycelium. Basidiospores 7.5–11.2 × 3.6–4.5 μm, ellipsoid to falciform, smooth, hyaline, inamyloid.

In summer to autumn, scattered on debris in forest edge. Distributed in northeast of China.

1152 *Gymnopus polyphyllus* (Peck) Halling

Pileus 3–6 cm in diam., convex when young, plano-convex to plane or slightly depressed at centre with age, reflexed at margin, sometimes undulate, glabrous, even, dry, dark brown at disc, pale pinkish brown at margin. Context thin, white. Lamellae adnate to decurrent, crowded, thin, sometimes furcate, white, entire at edge. Stipe 3–5 cm long, 2–5 mm thick, subcylindrical, hollow, cinnamon to white tinged with vinaceous brown, with bright grey tomentulum or strigose hairs, sometimes glabrous to pruinose at apex, densely tomentose to strigose at base, hollow. Odour garlic or foetid. Basidiospores 5.5–7.5 × 2.5–3.5 μm, smooth, hyaline, inamyloid.

Gregarious on ground in mixed forest. Distributed in northeast of China.

1153 *Hebeloma angustilamellatum* (Zhu L. Yang & Z.W. Ge) B.J. Rees

Pileus 3–10 cm broad, pale brown to yellow brown, becoming paler at margin, radially wrinkled, with minute velar remnants. Context white to dirty white. Lamellae close, narrow, pale yellow to brown. Stipe 5–12 cm long, 0.5–1.5 cm thick, cylindrical, whitish to pale brown. Annulus superior, tiny, fugacious. Basidiospores 9.5–11 × 7–8.5 μm, amygdaloid to subamygdaloid in side view, sublimoniformin in frontal view, smooth, ferruginous brown.

On ground in tropical to south subtropical forests. Distributed in Southern China.

1154 *Hebeloma crustuliniforme* (Bull.) Quél.

Pileus 3–7.5 cm broad, hemispherical to convex at first, then expanded, obtusely prominent; pale ochraceous when young, pale yellow at disc, becoming paler gradually to creamy white, greyish white or white at margin, then changing to deeper to creamy white, dark beige or grey beige, viscid when wet. Context thick, creamy white to white, with strong radish flavour, slightly bitterish. Lamellae subadnate, close, creamy white when young, pale earthy yellow to dark grey yellow when mature, white and irregularly denticulate at edges. Stipe 3–5 cm long, 1–1.5 cm thick, wide cylindrical, enlarged at base, with fibrillose squamules especially on upper portion, dirty white, gradually becoming pale yellow or pale yellow brown on lower portion. Basidiospores 10–13 × 6–7 μm, amygdaloid, verruculose, without a germ-pore, not dextrinoid.

Solitary or gregarious on ground in broad-leaf forests and coniferous forests. Poisonous. Distributed in Northeast and Northern China.

1155 *Hebeloma truncatum* (Schaeff.) P. Kumm.

Pileus 3–8 cm broad, conical to hemispherical when young, then gradually expanded to a convex to plane, often undulate at margin, sometimes umbonate at disc, glabrous, viscid when wet, paler as pale ochraceous when young, then changing to dark ochraceous to red brown. Context thin, white, with a radish odour, slightly bitterish. Lamellae adnate to nearly subsinuate, distant, pale earthy yellow to beige when young, then red brown. Stipe 3–5 cm long, 1–2 cm thick, cylindrical to subclavate, sometimes tapering toward base, solid, white, with white flocculent fibrils when young, then gradually changing to pale yellow or pale yellow brown on lower portion, with white fibrillose squamules. Basidiospores 7.3–10.3 × 4.9–5.9 μm, amygdaloid, slightly verruculose, without a germ-pore, pale yellow brown, dextrinoid.

Solitary or scattered on ground in mixed forests. Distributed in Central China.

This species is similar to a darker coloured species *H. sinapizans*, but the latter has a stronger radish flavour and slightly larger basidiospores.

1156 *Hemistropharia albocrenulata* (Peck) Jacobsson & E. Larss.

Pileus 3–12 cm broad, campanulate at first, then plane, often umbonate at disc, lustrous when dried, yellow brown to red brown, darker at disc, paler at margin, with erect or suberect flocculent scales; scales often dry and fugacious when mature; pileus margin glabrous and even, sometimes appendiculate with pale cinnamon flocculent velar remnants, revolute when mature. Context thick, white to dirty white. Lamellae slightly sinuate to adnate, pallid or dirty white at first, cinnamon brown when mature, distant, wide, thick, paler at edge, crenate. Stipe 3–10 cm long, 0.7–1.5 cm thick, cylindrical, subequal or slightly enlarged downwards, white or brownish and pruinose above annulus, cinnamon or dirty cinnamon and with fibrillose scales below annulus, soft to hollow insides. Annulus superior, fugacious. Basidiospores, 12–15 × 6.5–8 μm, subfusiform, smooth, brownish yellow.

In summer to autumn, gregarious or scattered, and sometimes subfasciculate on rhizosphere of broadleaf trees in mixed forests,. Edible. Distributed in Northeast China and Qinghai-Tibet Region.

1157 *Hohenbuehelia geogenia* (DC.) Singer

Pileus 3–10 cm broad, petaloid, flabelliform to spatulate, pale cinnamon to cinnamon at first, pale cinnamon to pale yellowish brown when mature, glabrous and undulate at margin, with white to grey white tomentulum in lower and middle portions. Context white, with smell of mushrooms. Lamellae white, decurrent, white to beige. Stipe 1–2 cm long, 0.5–1.5 cm thick, cylindrical, concolourous with pileus, with white to grey white tomentulum. Basidiospores 5.5–6.5 × 3.5–4 μm, elliptical, inamyloid.

Gregarious on ground in botanical gardens and parks. Distributed widely in China.

1158 *Hohenbuehelia petaloides* (Bull.) Schulzer

Pileus 3–7 cm broad, scoop-shaped, flabelliform to spatulate, tapering toward stipe, without posterior margin, white at first, then pale pink grey to pale brown, hygrophanous, subviscid, inconspicuously striate at margin. Context white, tasteless and odourless. Lamellae close, unequal, decurrent, narrow, white, pale cream or yellowish ochraceous brown when dried. Stipe 1–3 cm long, 0.5–1 cm thick, cylindrical, lateral, dirty white, tomentulose. Basidiospores 6–8 × 4–5 μm, subelliptical, smooth, hyaline, thin-walled, with refractive inclusion, inamyloid.

Gregarious to superposed or subfasciculate on rotten woods in mixed forests. Medicinal. Distributed in Northeast, Northern, Central, and Southern China.

1159 *Hohenbuehelia reniformis* (G. Mey.) Singer

Pileus 1–4 cm broad, semicircular to flabelliform, brown, to dark brown, densely with white, grey white to pale grey brown velutinous tomentum, subglabrous at margin, with denser tomentum at base, incurved or undulate at margin. Context thin, 2-layered, with a gelatinous upper layer and a white fleshy lower layer. Lamellae white to pale grey, pale yellow or beige when old, decurrent, narrow, moderately close. Stipe reduced, dorsally attached on lower surface of substrate, or as an attachment base, grey brown to pale yellow brown, white tomentose at base. Basidiospores 7.5–8 × 3–3.5 μm, cylindrical or elliptical, smooth, hyaline.

In summer to autumn, on rotten wood of various broadleaf trees such as *Ulmus pumila* and *Populus* sp. Edible. Distributed in Northeast, Northwest, Northern, and Central China.

1160 *Hydropus atriceps* (Murrill) Singer

Pileus 0.8–1.8 cm broad, campanulate to convex when young, glabrous, gradually expanded to plane at maturity, finely fibrillose, blackish brown, chestnut brown to black brown, thin and estriate at margin. Context dark brown to nearly brown; odour unobvious or lacking. Lamellae 2.5–3 mm broad, dark grey to grey brown, moderately close, unequal, sinuate or subadnate, darker at edge. Stipe 3.5–4.7 cm long, 2–5 mm thick, cylindrical, often curved, dark grey to black, glabrous, fibrous, hollow. Basidiospores 6–7.5 × 4–5 µm, elliptical, smooth, hyaline.

In autumn, fasciculate on rotten wood. Distributed in Northeast China.

1161 *Hydropus floccipes* (Fr.) Singer

Pileus 1.2–2 cm broad, convex when young, plane at maturity, glabrous, hygrophanous, slightly umbonate at disc, brown to dark brown; becoming paler to pale brown at margin, striate. Lamellae sinuate, dirty white, hygrophanous, distant or moderately close. Context thin, dirty white. Stipe 2.5–4 cm long, 2–3.5 mm thick, hollow, dirty white at apex, fibrillose, becoming brown or deep red brown downward the base, gradually becoming glabrous downwards, sericeous. Basidiospores 7–10 × 5–6 µm, elliptical, smooth, hyaline.

In summer, on barks or trunks. Distributed in Northern China.

1162 *Hydropus marginellus* (Pers.) Singer

Pileus 0.6–1.5 cm broad, campanulate to convex when young, gradually becoming plane at maturity, sometimes subinfundibuliform, convex at disc, glabrous, translucently striate; dark when young, smoky brown, pale brown to grey brown when mature; narrowly incurved at margin. Context membranous; odour unobvious. Lamellae white, adnate to subdecurrent, brown at edge, changing to black when old. Stipe 1–2 cm long, 3–7 mm thick, cylindrical, often curved, greyish brown, glabrous, finely white pruinose, hollow. Basidiospores 5.5–7.5 × 3–5 µm, elliptical, smooth, hyaline.

In summer to autumn, on trunks of *Abies fabri*. Distributed in Northeast China.

1163 *Hydropus nigrita* (Berk. & M.A. Curtis) Singer

Pileus 1–3 cm broad, hemispherical to campanulate when young, convex to plane at maturity, with a margin exceeding lamellae, grey to dark grey, changing to black when mature or touched. Context white when young, changing to black when old or injured. Lamellae sinuate, close, white at first, changing to black when mature or touched. Stipe 1.5–3.5 cm long, 2–4 mm thick, cylindrical or flattened-cylindrical, with grey white tomentum, white to greyish white when young, becoming black when old, solid to hollow, fibrous. Basidiospores 4.5–5.5 × 3–5 µm, subglobose, smooth, hyaline, inamyloid.

In summer, gregarious on rotten woods of coniferous trees. Distributed in Northeast China.

The main characteristic of this species is that the fruit body turns black when old or touched by hands, easily to be identified.

1164 *Hygrocybe cantharellus* (Schwein.) Murrill

Pileus 2–4 cm broad, obtusely conical to convex when young, then depressed at centre to infundibuliform; sericeous at first, then squamulose at disc, incurved or undulate at margin, red brown to orange red when young, fading when old. Context thin, dirty white to orange yellow. Lamellae decurrent, subdistant, orange to yellow, even at edge. Stipe 4–7 cm long, 3–5 mm thick, cylindrical or slightly flattened-cylindrical, subequal, brittle, glabrous, orange yellow at upper portion, white at base, solid at first, then hollow. Basidiospores 7.5–11 × 5–6.5 µm, elliptical, smooth, hyaline.

In summer, gregarious or scattered on ground in Korean pine and spruce-fir forests. Edible. Distributed in Northeast, Central, and Qinghai-Tibet Regions, China.

1165 *Hygrocybe conica* (Schaeff.) P. Kumm.

≡ *Hygrophorus conicus* (Schaeff.) Fr.

Pileus 2–7 cm broad, conical at first, then gradually expanded, sharply acute at centre, glutinous, becoming dry quickly, with an epicutis often rimose into fibrillose **hairs**, often lacerated and reflexed at margin, red brown or orange yellow when young, paler at margin, olive grey to black when mature, changing quickly to black when injured. Context thin, pale red brown at first, gradually changing to grey black, changing to black when injured. Lamellae free, close, thin, dirty white to orange yellow, black at age, often serrulate at edge. Stipe 6–13 cm long, 0.5–1.2 cm thick, hollow, cylindrical, often curved, wet or dry, not viscid, very brittle, dark red or orange yellow at upper portion, dirty white at base, changing to black when injured and when mature. Basidiospores 10.5–12.9 × 5.6–7.8 µm, elliptical, smooth, hyaline.

In summer to autumn, gregarious or scattered on ground in forests. Poisonous. Distributed in Northeast, Northern, and Central China.

1166 *Hogrocybe* cf. *cuspidata* (Peck) Murrill

Pileus 2–5 cm broad, conical at first, gradually becoming plane, glabrous, viscid, scarlet; margin orange yellow to yellow, striate, reflexed or undulate when mature, often lacerated. Context thin, pale yellow or white, slightly changing to black or inconspicuously greyish when injured. Lamellae free or subadnexed, subdistant at maturity, white to pale yellow; edge brittle and often impaired. Stipe 5–9 cm long, 0.5–1 cm thick, subcylindrical, glabrous or longitudinally grooved, hollow, orange at upper portion, gradually becoming yellow downwards, whitish at base. Basidiospores 7–10 × 4–7 μm, elliptical, smooth, hyaline.

Solitary or scattered on open fields in broad-leaf forests. Distributed in Southern China.

1167 *Hygrocybe firma* (Berk. & Broome) Singer

Pileus 2–4 cm broad, plane to flat at first, slightly depressed at centre when mature, deep red, red to orange yellow. Context thin, white with brownish tint. Lamellae decurrent, white at first, pale yellow when mature. Stipe 3–5 cm long, 3–5 mm thick, cylindrical, yellow, sometimes reddish. Basidia dimorphous; lager basidia 50–60 × 14–16 μm; smaller basidia 30–40 × 6–8 μm. Basidiospores dimorphous; larger basidiospores 13–18 × 7.5–11 μm, elliptical to wide elliptical; smaller basidiospores 6–9 × 3.5–5 μm, elliptical, smooth, hyaline.

In summer to autumn, on soil slopes of roadside or ground in tropical forests. Distributed in Southern China.

1168 *Hygrocybe griseobrunnea* T.H. Li & C.Q. Wang

Pileus 1.6–3.2 cm broad, plane when young, reflexed at margin when mature, often perforated at centre, grey, partially with violet tinge, with dark brown squamules, some squamules easily falling off when old. Context thin, white to concolourous with pileus. Lamellae 3 mm broad, whitish to white, adnexed to short decurrent, translucent, waxy, fragile, approximately with four complete lamellae per cm, with 1–3 lamellulae between 2 complete lamellae. Stipe 1.2–2.8 cm long, 3–6.5 mm thick, cylindrical, white to pale violet or grey violaceous, glabrous. Basidiospores 6–8.5 × 4–6.5 µm, elliptical, smooth, hyaline.

Gregarious on sandy ground in broad-leaf forests. Distributed in Southern China.

1169 *Hygrocybe miniata* (Fr.) P. Kumm.

≡ *Hygrophorus miniatus* (Fr.) Fr.

Pileus 1–4 cm broad, oblate hemispherical to obtusely conical at first, then becoming plane, slightly umbonate at centre, not viscid, subglabrous or minutely squamulose, reddish brown when moist, paler when dried. Context thin, pale yellow. Lamellae adnexed to subdecurrent, distant, thick, waxy, pale yellow. Stipe 3–5 cm long, 3–5 mm thick, cylindrical or slightly flattened, sometimes curved, solid at first, then hollow, cartilaginous, glabrous, orange and slightly tincted with light red brown on upper portion, paler on lower portion. Basidiospores 7.5–11 × 5–6 µm, elliptical, smooth, hyaline.

In late spring to autumn, scattered to gregarious on grassy ground in broad-leaf forests. Edible. Distributed in Northeast, Central, and Southern China.

1170 *Hygrocybe* ef. *nigrescens* (Quél.) Kühner

Pileus 2–6 cm broad, conical at first, then bamboo-hat-shaped, orange red, orange yellow or bright red, often with dark striae, changing to black when injured or dried, often lacerated at margin. Context pale yellow, changing to black when injured. Lamellae pale yellow. Stipe 4–12 cm long, 0.5–1.2 cm thick, cylindrical, orangish, longitudinally striate, gradually becoming hollow insides, changing to black when injured, more easily to black at base. Basidiospores 10–12 × 7.5–8.5 µm, elliptical, smooth, subhyaline to yellowish.

In summer to autumn, gregarious or scattered on ground in coniferous or broad-leaf forests. Poisonous according to records. Patients, has a long incubation period after poisoning, causing severe vomiting and diarrhea, and even causing shock to death due to dehydration. Distributed widely in China.

Molecular evidences show that this Chinese species is somewhat different from the European species *H. nigrescens*. It is now treated as a reference species here.

In summer and autumn, on ground in mixed forests. Edible. Distributed in Northeast and Southern China.

1171 *Hygrocybe psittacina* (Sehaeff.) P. Kumm.

≡ *Hygrophorus psittacinus* (Schaeff.) Fr.

Pileus 0.5–3.5 cm broad, initially bamboo-hat-shaped or conical-umbonate, then gradually becoming plane, green and glutinous at first, fading to yellow or orange yellow when old or dried. Context thin, yellowish. Lamellae adnate, subdistant, unequal, green at first, then fading to orange yellow. Stipe 1.5–5 cm long, 1–4 mm thick, yellow green at first, then yellow or orange. Basidiospores 7–9 × 4.5–5 μm, elliptical, smooth, hyaline to yellowish.

1172 *Hygrophoropsis aurantiaca* (Wulfen) Maire

Pileus 2–7 cm in diam., plane, orange-red to yellow brown, darker at disc, with concolourous velvety squamules. Context pale yellow. Lamellae decurrent, close, narrow, interveined, orange yellow to orange red, blunt at edge. Stipe 3–6 cm long, 3–8 mm thick, cylindrical, brownish yellow. Basidiospores 6–8 × 4–5.5 μm, elliptical to long elliptical, smooth, hyaline to yellowish.

In summer and autumn, on ground in forests. Distributed in most regions of China.

1173 *Hygrophorus cossus* (Sowerby) Fr.

Pileus 2–4 cm in diam., hemispherical, oblate hemispherical, convex to plane, white when young, pale yellow at maturity, viscid, slightly incurved at margin. Context white when young, becoming yellow when aged, with an obvious smell of carpenter moth. Lamellae adnexed to short decurrent, white to pale luteotestaceous, subdistant, with 1–3 small lamellulae between complete lamellae. Stipe 3–7 cm long, 0.5–1 cm thick, cylindrical, usually thinner at base, white, fibrillose, viscid. Basidiospores 6–7 × 4–5 μm, elliptical, smooth, hyaline.

Solitary or gregarious on ground in broad-leaved forests. Distributed in central and southern China.

1174 *Hygrophorus lucorum* Kalchbr.

Pileus 3–5 cm in diam., subhemispherical when young, then expanded gradually, slightly protruding at centre, glutinous when moist, glabrous, lemon yellow. Context dirty white or pale yellow, slightly thicker at centre, with a slight odour and a mild taste. Lamellae decurrent, subdistant, initially dirty white, gradually becoming to pale yellow. Stipe 5–9 cm long, 0.7–1.5 cm thick, cylindrical, subequal or slightly enlarged downwards, white or pale yellow, viscid, solid at first, hollow when old. Basidiospores 7.5–9 × 4–6.5 μm, elliptical, smooth, hyaline.

In summer to autumn, scattered or gregarious on ground in coniferous-broadleaf mixed forests. Edible. Distributed in Northeast China, Central China and Inner Mongolia region.

1175 *Hygrophorus pudorinus* (Fr.) Fr.

Pileus 5–13 cm in diam., initially flat hemispherical, then expanded, initially incurved at margin, with white fibrils, fleshy, glabrous, subviscid when moist, initially pale yellow, gradually pink to pinkish red, darker at disc. Context thick, white, taste mild, with odour of rosin. Lamellae adnate to subdecurrent, distant, thick, unequal, interveined, white, gradually becoming pinkish, occasionally with pink spots. Stipe 5–12 cm long, 2–3 cm thick, cylindrical, subequal or thinner at base, with white villous squamules or flocculent granules at upper portion, fibrillose on lower part, concolourous with pileus, slightly viscid when moist, solid. Basidiospores 8.5–11.2 × 5.3–6.2 μm, elliptical, smooth, hyaline.

In summer and autumn, gregarious on ground in coniferous and broadleaf mixed forests. Edible. Distributed in Northeast China, Central China and other regions.

1176 *Hygrophorus russula* (Schaeff.) Kauffman

Pileus 6–15 cm in diam., convex at first, then gradually expanded to plane, sometimes depressed at disc, slimy when fresh, but easily dry, glabrous, sometimes minutely squamulose, inrolled at margin when young, pale red to pale pink, often striate and weakly speckled. Context thick, tough, white to pink. Lamellae adnate to decurrent, close, white at first, with reddish spots when old, unequal, waxy. Stipe 4–10 cm long, 1.2–3.5 cm thick, cylindrical, subequal, white at first, nearly concolourous with pileus when old, finely striate, solid. Basidiospores 5.5–8 × 3.3–4.5 µm, elliptical, smooth, subhyaline to yellowish.

In autumn, gregarious on ground in mixed coniferous and broadleaf forests. Distributed in Northeast China, Central China and other regions.

1177 *Hypholoma capnoides* (Fr.) P. Kumm.

Pileus 2–4 cm broad, hemispherical at first, then convex to plane, incurved at margin at first, then expanded to sometimes reflexed, subhygrophanous when wet, red brown to ochraceous brown or pale orange brown; margin tincted with grey yellow to grey white, with velar remnants, connected by a white cortinoid veil to stipe when young; cortina evanescent at maturity. Context white to grey. Lamellae adnate to sinuate, white to smoky violaceous brown, and finally deep grape purple brown. Stipe 3–8 cm long, 2–7 mm thick, cylindrical, white to yellow white at upper portion at first, gradually changing to brown to ferruginous brown from base at maturity, with an annular vestige. Basidiospores 7–8 × 4.5–5 µm, elliptical to slightly elliptical, smooth, pale purplish brown or purplish grey.

In summer to autumn, caespitose to fasciculate on rotten woods of conifers or mixed forests. Distributed in Northeast, Northwest, and Central China.

1178 *Hypholoma fasciculare* (Huds.) P. Kumm.

≡ *Naematoloma fasciculare* (Huds.) P. Karst.

Pileus 0.3–4 cm broad, conical to campanulate at first, subhemispherical to plane, obtuse to slightly acute at centre, sulfureous yellow to slightly red brown to orange brown at apex, glabrous, sulfureous yellow to greyish sulfureous yellow at edge, absorbing water to subhygrophanous when wet, easily changing to dark brown to dark reddish brown when dried, or dark brown in hygrophanous areas, sometimes unchanging when dried, appendiculate with yellow cortinoid velar remnants at margin at first; velar remnants evanescent at later stage. Context pale yellow to lemon yellow. Lamellae sinuate, sulfur yellow at first, then gradually changing to olivaceous green, finally changing to olivaceous violaceous brown. Stipe 1–5 cm long, 1–4 mm thick, cylindrical, sulfureous yellow, gradually changing to orange yellow to dark reddish brown downwards, sometimes with velar vestiges or a fugacious annulus, with yellow tomentum at base. Basidiospores 5.5–6.5 × 4–4.5 µm, elliptical to long ellipsoid, smooth, lavender grey.

In summer to autumn, caespitose to fasciculate on logs, fallen wood, decayed branches of coniferous and broad-leaved trees, or on decayed wood buried in the ground. Poisonous. Distributed widely in China.

1179 *Hypholoma lateritium* (Schaeff.) P. Kumm.

= *Naematoloma sublateritium* (Schaeff) P. Karst.

Pileus 1–9 cm broad, hemispherical to convex at first, then becoming broadly convex to plane, sometimes umbonate-globose, slightly incurved or reflexed when mature at margin, not viscid to moist, pale umbrinous or reddish brown to brick red, lighter at margin, white, yellowish white, greyish yellow to pale yellow at first, covered with white to greyish white woolly flocculence, or with fugacious velar remnants; often irregularly cracked at centre. Context white to whitish, becoming darker when injured. Lamellae sinuate to subadnate, white to yellowish white at first, gradually changing to olive green and then to dark grey, and finally to pale violaceous brown to deep violaceous brown, sometimes olivaceous brown or olivaceous yellow. Stipe 3–10 cm long, 4–8 mm thick, white to yellowish white, and hygrophanous white at upper portion, brown to ferruginous brown at lower portion. Annulus absent. Basidiospores 6–7 × 4–5 µm, wide elliptical to elliptical, smooth, lilac grey.

In late summer and autumn, fasciculate to caespitose on bark, lumbers, fallen logs, stumps of broad-leaved trees, or on decayed wood buried in ground. Poisonous. Distributed in Northeast, Northern, Central, Inner Mongolia, and Northwest China.

1180 *Hypsizygus marmoreus* (Peck) H.E. Bigelow

Pileus 2–5 cm broad, flat hemispherical when young, then slightly plane, slightly prominent at disc, dirty white, greyish white to yellow when fresh, glabrous, hygrophanous, with looming pale brown macules at centre as a mosaic pattern, grey brown when dried, azonate, rough, acute at margin, incurved when dried. Context rather thick, white. Lamellae subadnate, dirty white, changing to pale yellow brown when dried, close or subdistant, unequal, brittle. Stipe 3–11 cm long, 0.5–1 cm thick, cylindrical, slender and slightly curved, white, glabrous or longitudinally striate, solid, often caespitose and confluent or branched at base. Basidiospores 4–5.5 × 3.5–4.5 μm, wide elliptical to subglobose, smooth, hyaline.

In late summer to autumn, caespitose on dead wood and fallen wood of broad-leaved trees. Edible, cultivatable. It is an excellent edible fungus, with commercial names of "crab-flavored mushroom" and "seafood mushroom". Distributed in Northeast China and Qinghai-Tibet Regions.

1181 *Hypsizygus ulmarius* (Bull.) Redhead

Pileus 8–16 cm broad, hemispherical to flat hemispherical at first, gradually plane, glabrous, light ochraceous at centre, sometimes cracked, pale yellow at margin. Context thick, white. Lamellae wide, sinuate, close, white to whitish. Stipe 5–9.5 cm long, 1–2 cm thick, cylindrical, solid, slightly curved, subeccentric, white. Basidiospores 5.4–6.5 μm, globose to subglobose, smooth, hyaline.

In summer to autumn, subcaespitose or caespitose on trunks of *Ulmus pumila* or other broad-leaved trees. Edible. Distributed in Northeast and Northwest China.

1182 *Inocybe adaequata* (Britzelm.) Sacc.

Pileus 2.2–6 cm broad, conical to campanulate when young, and nearly plane to reflexed at margin when mature, reddish brown to vinaceous red, then becoming darker, fibrillose, rimulose, obtusely prominent at disc, incurved at margin when young, and then expanded. Context fleshy, pink. Lamellae adnate to subfree, close, pale pink when young, brown when mature, serrate at margin. Stipe 9–11 cm long, 0.9–1.2 cm thick, cylindrical, equal, fibrillose, dark pink or vinaceous, blackening when mature or injured, longitudinally striate, solid. Basidiospores 9.5–14.5 × 6–7.5 μm, subphaseoliform, smooth, yellowish brown.

In summer, solitary or scattered on ground in bamboo groves, broad-leaved or coniferous forests. Distributed in Northeast, Northwest, Qinghai-Tibet Regions, China.

1183 *Inocybe assimilata* Britzelm.

Pileus 1.5–2 cm broad, campanulate to hemispherical when young, then bamboo hat-shaped or plane, with an obvious obtuse umbo at disc, sometimes reflexed at margin, dark brown to pale brown, uniformly coloured, coarsely fibrillose, ramulose to lacerate at margin. Context fleshy, white. Lamellae close, adnate, milky white at first, then olive grey, pale brown at maturity, paler at edge, narrower near stipe, becoming broader gradually toward pileal margin. Stipe 3–4 cm long, 2–3 mm thick, cylindrical, equal or slightly becoming thicker downwards, enlarged and marginate at base, up to 5 mm thick at enlarged part, pale brown to with pale fleshy tinge, gradually lighter on middle and lower portions, solid. Basidiospores 7–9 × 5–6.5 μm, irregularly rectangular, inconspicuously verrucolose, occasionally obviously prominently verruculose, pale brown to brown.

In summer or autumn, scattered in broad-leaved or coniferous forests. Distributed in Northeast and Central China.

1184 *Inocybe asterospora* Quél.

Pileus 2–3.5 cm broad, earthy yellow, obviously rimulose, radially striate, lacerate at margin, prominent at centre, with inconspicuous appressed scales at umbo, without cortinoid velar remnants at margin. Context with a strong earthy smell, fleshy, white. Lamellae up to 3.5 mm broad, sinuate or subfree, white at first, then grey, medium close, thin, whitish at edge. Stipe 6–8 cm long, 3–5 cm thick, cylindrical, solid, concolourous with pileus, gradually thicker downwards, densely pruinose to base, bulbous and marginate at base, 6 mm thick at enlarged portion. Basidiospores 10–11 × 8–9.5 μm, stellate, pale brown.

In summer to autumn, solitary on ground in broad-leaved forests. Distributed in Northeast and Northwest China.

The main characteristic of this species is the stellate basidiospores, which are similar to those of *I. miyiensis*, but the latter has larger basidiospores and basidia mostly with only two sterigmata.

1185 *Inocybe caroticolor* T. Bau & Y.G. Fan

Pileus 1.7–3.3 cm broad, conical to campanulate when young, bamboo hat-shaped to plane when mature, obviously obtusely umbonate at centre, with radially appressed scales, fibrillose, lacerate at margin, orange yellow to apricot yellow; scales concolourous with pileus when young, gradually turning to brown to reddish brown when mature; background orange yellow to apricot yellow or ochraceous yellow. Context obviously aromatic, white to light apricot yellow. Lamellae up to 3 mm broad, adnate, close, pale orange yellow to apricot yellow when young, dark apricot yellow to brown when mature, concolourous and even at edge. Stipe 3–4.2 cm long, 2–3 mm thick, cylindrical, equal, solid, pale orange yellow to apricot yellow, fully pruinose-granular. Basidiospores 6.5–9 × 5–6 μm, with 7–9 obvious to inconspicuous nodules, yellowish brown.

In summer to autumn, solitary or scattered on the roadside in *Quercus variabilis*. Distributed in Central China.

1186 *Inocybe casimiri* Velen.

Pileus 1–2 cm broad, hemispherical when young, densely with fine bristle-like scales, brownish black, gradually plane when mature, with scales erect at centre and gradually appressed toward margin, dark brown, with fugacious cortinoid partial veil or velar remnants at margin. Context white, fleshy, brownish at pileipellis. Lamellae adnate or sinuate, moderately close, greyish white when young, then pale brown to brown, lighter at edge. Stipe 2.8–3.5 cm long, 2–2.5 mm thick, solid, equal, whitish with brownish tint above velar remnants, woolly below, brown and gradually darker downwards. Basidiospores 10–12.5 × 8–9.5 μm, with 13–15 nodules, pale brown. Pleurocystidia absent. Cheilocystidia 29–48 × 11–15 μm, thin-walled, hyaline.

In autumn, on heavily brown rotten stumps. Distributed in Northeast China and Qinghai-Tibet Regions, China.

This species is similar to *I. lanuginosa*, however the latter species has thick-walled pleurocystidia.

1187 *Inocybe corydalina* Quél.

Pileus 2.6–4.5 cm broad, campanulate when young, convex to plane when mature, obtusely umbonate at centre, with a margin exceeding lamellae, with appressed scales, fibrillose, greyish brown to brown, sometimes ochraceous brown, atrovirens at centre, gradually lighter toward margin, sometimes inconspicuously so. Context 4–6 mm thick, fleshy, white to dirty white, fragrant. Lamellae adnate, close, thin, white to greyish white when young, brownish when mature, slightly serrulate and paler at edge. Stipe 6–9.5 cm long, 4–6 mm thick, cylindrical, equal; slightly enlarged at base, 7–8 mm thick, solid, grey to dark grey; with inconspicuous white to greyish white fibrous scales at apex, fibrillose downward, longitudinally striate, often spiral; dark green or atrovirens at centre and lower portions, sometimes inconspicuously so; with greyish white tomentose mycelium at base. Basidiospores 8–10 × 5–6 μm, elliptical to amygdaliform, smooth, yellow to yellowish brown.

In summer, scattered on ground in mixed forests with broad-leaved trees and *Pinus koraiensis*. Distributed in Northeast China.

1188 *Inocybe curvipes* P. Karst.

Pileus 2.2–3.6 cm broad, conical when young, gradually plane when mature, obviously umbonate at centre, smoky brown at prominent region, gradually lighter to margin, with appressed radially scales, lacerate at margin with age. Context up to 3 mm thick, white, with a slight earthy smell. Lamellae up to 3.5 mm broad, adnate, close, unequal, greyish white with olivaceous tinct when young, brown when mature, uneven and slightly lighter at edge. Stipe 3.5–4.5 cm long, 3–5 mm thick, cylindrical, equal, tobacco-brown, lighter at apex and base, with tomentose-fibrillose squamules, with white mycelium at base, solid. Basidiospores 9–11.5 × 5–6 μm, projectile-shaped, obviously to inconspicuously nodulose, pale brown.

In summer, solitary or scattered on ground in forests or on roadside of forest edge, with close relationships of *Populus*, *Salix*, or *Larix gmelinii*. Distributed in Northeast China.

The pileus and stipe of this species are similar to those of *I. lanuginosa*, however the latter species has darker colour and obviously different shape of basidiospores.

1189 *Inocybe dulcamara* (Pers.) P. Kumm.

Pileus 1.5–2.5 cm broad, hemispherical when young, nearly plane to centrally depressed when mature, densely with fine squamules radially spreading from centre to margin, appendiculate with cortinoid velar remnants at margin when young, brownish yellow, darker at centre, gradually lighter toward margin. Context 0.3 cm thick, fleshy, earthy yellow, without a conspicuous odour. Lamellae up to 3 mm broad, decurrent, yellowish brown with olive tint, medium close, finely serrulate at edge. Stipe 2.2–3 cm long, 3–4 mm thick, cylindrical, equal, fibrillose, pruinose to finely furfuraceous-granular at apex. Basidiospores 8–10.5 × 6–7 μm, elliptical to subphaseoliform, smooth, yellowish brown.

In summer to autumn, solitary to scattered in broad-leaved forests or on roadside. Poisonous. Distributed in Northeast, Northern, Northwest, Qinghai-Tibet Regions, China.

1190 *Inocybe erubescens* A. Blytt

Pileus 3–6.5 cm broad, conical to campanulate when young, bamboo hat shaped to plane when mature, decurved when young, obviously to not obviously incurved at margin, reflexed strongly at margin with age, umbonate or prominent at centre, dry, fibrillose, rough, finely rimulose, sometimes lacerate at margin when mature; with cortinoid velar remnants at margin when young; straw yellow to ochraceous yellow, gradually pinkish to with orange red tint when injured or age. Context with an inconspicuous farinaceous taste, fleshy, white to with pink or orange red tint. Lamellae close, adnate, narrow, dirty white to greyish white when young, pinkish when mature or injured, concolourous or lighter at edge. Stipe 6.5–9 cm long, 0.6–1.2 cm thick, cylindrical, equal or tapering at apex, bulbous at base, solid, finely fibrillose, roughly with fibrillose or furfuraceous squamules at apex, with white mycelium on middle and lower portions, white to dirty white, gradually with pink or orange red tint when mature. Basidiospores 11–13.5 × 6.5–7.5 μm, elliptical to long ellipsoid, obtuse at apex, smooth, yellowish brown.

In summer, solitary or scattered on ground in Fagaceae forests. Distributed in Northern, Central, Northwest China.

This species is similar to *I. godeyi*, however the latter species has white pruinose granules on stipe. This species is reported as a more poisonous species within *Inocybe*, but is still eaten by some folks around Beijing.

1191 *Inocybe geophylla* var. *geophylla* (Bull.) P. Kumm.

Pileus 1.1–1.5 cm broad, conical when young, then gradually plane, obviously umbonate at centre, glabrous, sericeous, finely rimulose to obviously rimose at margin when mature, white or yellowish. Context with a strong earthy smell, fleshy, white or yellowish. Lamellae up to 2 mm broad, white when young, then grey to pale brown, subdistant, adnate, lighter at edge. Stipe 3–5.5 cm long, 2–2.5 mm thick, cylindrical, equal, slightly thicker at base, white, with pruinose squamules at apex, fibrillose in lower portion, solid, often infested by insects. Basidiospores mostly 8.5–9.5 × 5–6 μm, rarely up to 12.5 × 6.5 μm, elliptical, smooth, pale brown.

In summer to autumn, solitary or scattered on ground in broad-leaved or coniferous forests. Distributed in Northeast, Inner Mongolia and Northwest China.

This species is one of the few white species within *Inocybe*, and also is a relatively common species. It is similar to *I. umbratica*, however the latter species has white pruinose granules all over the stipe.

1192 *Inocybe geophylla* var. *lilacina* (Peck) Gillet

Pileus 1.8–2.7 cm broad, conical or campanulate when young, nearly plane when mature, with an acute prominence at centre, glabrous and fibrillose, light lilac, darker when young, light earthy yellow or beige yellow at prominence. Context with an earthy smell, fleshy, light violaceous at pileipellis. Lamellae close, unequal, sinuate and slightly decurrent, lilac when young, then brownish grey to ferruginous brown, uneven and lighter at edge. Stipe 4.5–6.5 cm long, 2–4.5 mm thick, cylindrical, slightly thinner at centre, gradually thicker downwards, obtuse or slightly enlarged at base, light earthy yellow or beige yellow, solid, whitish or violaceous light violaceous. Basidiospores 9–10 × 5–6 µm, elliptical to reniform, obtuse at apex, pale brown.

Autumn, scattered on ground in spruce-fir forests, mixed forests or broad-leaved forests. Distributed in Northeast, Qinghai-Tibet Regions, China.

This species is one of the easily recognizable species within *Inocybe* for its pale lilac pileus. Although it is very similar to *I. geophylla* var. *violacea*, the latter species has smaller and blueish violaceous fruiting bodies.

1193 *Inocybe geophylla* var. *violacea* (Pat.) Sacc.

Pileus 1–1.5 cm broad, conical when young, then gradually plane, with a small prominence or umbo at centre, blueish violaceous to dark lilac, yellow at central prominence, glabrous. Context 1–2 mm thick, with a weak earthy smell, fleshy, white. Lamellae moderately close, greyish white to yellowish brown, adnate, uneven at edge. Stipe 2.2–3.2 cm long, 1.5–2 mm thick, cylindrical, equal or gradually thicker downwards, enlarged at base, blueish violaceous, pale yellow at enlarged region, with pruinose squamules at apex, solid. Basidiospores 8.5–10 × 5–6 µm, subamygdaliform, slightly obtuse at apex, smooth, yellowish brown.

In autumn, scattered on ground in broad-leaved forests. Distributed in Northeast China.

This species is similar to *I. geophylla* var. *lilacina*, however the latter species has larger and lighter lilac fruiting bodies.

1194 *Inocybe glabrescens* Velen.

Pileus 2.5–3 cm broad, campanulate when young, then hemispherical, plane at centre, without an umbo, fibrillose, beige yellow to straw yellow, pale brown at centre, lighter and lacerate at margin. Context without an obvious odour, white, fleshy, tinged with brown at pileipellis. Lamellae close, adnate or sinuate, unequal, thin, white when young, then ivory, finely denticulate at edge. Stipe 5–7 cm long, 6–7 mm thick, cylindrical, solid, often infested by insects, equal or slightly thicker at base, with white pruinose granules at apex, gradually with less granules downwards, longitudinally striate, subglabrous, dirty white to milky white, with white tomentose mycelium at base when young. Basidiospores 8.5–9.5 × 4.5–5.5 µm, amygdaliform, slightly pointed at apex, smooth, brown.

In autumn, on ground in broad-leaved forests with trees such as *Quercus mongolica*, *Acer*, and *Phellodendron amurense*. Distributed in Northeast China.

1195 *Inocybe godeyi* Gillet

Pileus 1.8–4.2 cm broad, campanulate when young, then bamboo hat shaped to plane; incurved at margin when young, then plane, with an obviously obtuse umbo at centre, sericeous, occasionally with inconspicuous squamules, pale brown, lighter at margin, gradually with orange red to pink tint when mature, changing to orange red to pink when injured. Context with an earthy smell, fleshy, white when young, tinged with orange red when mature. Lamellae 2–4 mm broad, adnate, close, white to greyish white when young, gradually with orange red to brick red tint when mature or injured, lighter at edge. Stipe 3.7–5.8 cm long, 4–6 mm thick, cylindrical, equal, solid, sericeous, longitudinally striate, often with white pruinose granules, beige yellow to pale flesh brown when young, then gradually changing to orange red, bulbous and obviously marginate at base. Basidiospores 8.5–11 × 5.5–7 μm, amygdaliform, apiculate, smooth, yellowish brown.

In summer to autumn, on ground in broad-leaved forests. Distributed in Central and Northwest China.

This species is easily recognized because it turns red when mature or injured, the difference from *I. erubescens* is that the latter species has no pleurocystidia.

1196 *Inocybe grammata* Quél.

Pileus 2.2–3.5 cm broad, campanulate when young, gradually plane when mature, obviously obtusely prominent at centre, fleshy pink to pinkish brown, glabrous, dry, not rimose. Context with a strong earthy smell, fleshy, pinkish white. Lamellae up to 5 mm broad, greyish white at first, then changing to greyish brown with fleshy pink tint, close, adnate, gradually narrower toward stipe, thin, uneven or slightly serrulate at edge, with a complete edge. Stipe 4–5.5 cm long, 3–4 mm thick, cylindrical, equal, bulbously enlarged up to 6 mm thick at base, solid, fleshy pink, pruinose. Basidiospores 7.5–9 × 5–6 μm, angular to verruculose, yellowish brown.

In autumn, solitary on ground in broad-leaved or coniferous forests and on roadside. Distributed in Northeast and Central China.

1197 *Inocybe hainanensis* T. Bau & Y.G. Fan

Pileus 1.8–2.2 cm broad, conical to bamboo-hat shaped or subhemispherical when young, convex to nearly plane when mature, fibrillose, with an epicutis easily cracked into inconspicuous patchy scales on the prominent area when mature, with fugacious cortinoid velar remnants at margin, straw yellow to brownish. Context 2 mm thick, milky white to pale yellow. Lamellae adnate, close, brownish yellow to brown. Stipe 2–2.5 cm long, 2–3 mm thick, cylindrical, equal, solid, slightly or inconspicuously enlarged at base, light lilac. Basidiospores 8–10 × 5–6 μm, yellowish brown, long ellipsoid to slightly angular.

On ground in tropical broad-leaved forests. Distributed in Southern China.

1198 *Inocybe hirtella* Bres.

Pileus 1.5–2 cm broad, hemispherical when young, gradually plane when mature, with an obtuse prominence at centre, sometimes inconspicuously so, pendent to plane and sometimes lacerate at margin, fibrillose to tomentose, earthy yellow to ochraceous yellow, with orangish or inconspicuous darker squamules at prominence. Context with a distinct flavour of bitter almond to earthy smell, fleshy, white. Lamellae 2–3 mm broad, adnate, close, white to greyish white when young, brown when mature, lighter at edge. Stipe 3.3–4.5 cm long, 3–4 mm thick, cylindrical, equal, solid, enlarged or inconspicuously so at base, longitudinally striate, fleshy pink, with white pruinose granules, with white tomentose mycelium at base. Basidiospores 8.5–9.5 × 5–6 μm, submygdaliform to elliptical, obtuse to slightly acute at apex, smooth, yellowish brown.

In summer to autumn, scattered on ground in broad-leaved forests. Distributed in Central and Northeast China.

This species is close to *I. muricellata*, however, the latter species has relatively larger fruiting bodies and a yellow to bright yellow stipe.

1199 *Inocybe lacera* (Fr.) P. Kumm.

Pileus 1–1.5 cm broad, conical when young, then campanulate, with a conspicuous or inconspicuous prominence at centre, brown to dark brown, gradually lighter to margin, rough to densely with brown squamules. Context fleshy, white, with brown tinge at pileipellis. Lamellae up to 3 mm broad, moderately close, adnate, greyish white when young, gradually changing to yellowish brown when mature, lighter at edge. Stipe 3–3.5 cm long, 1–1.5 mm thick, cylindrical, equal or gradually thicker to base, 2 mm thick at enlarged region, fibrillose, milky white to greyish white at apex and upper portion, becoming brownish grey downwards, solid. Basidiospores 10.5–13 × 4.5–5.5 μm, long ellipsoid to long oblong, slender, occasionally weakly angular, obtuse or slightly truncated at apex, smooth, yellowish brown.

In summer to autumn, solitary or scattered on ground and roadside in broad-leaved or coniferous forests. Distributed in Northeast, Northwest and Central China.

1200 *Inocybe lanuginosa* (Bull.) P. Kumm.

Pileus 0.8–1.5 cm broad, hemispherical when young, bamboo hat shaped when mature, covered with dark brown bristle-like scales, with an inconspicuous prominence at centre, with erect scales at centre, gradually with radially appressed scales to margin, lighter when young. Context milky white with brownish tint, without special odour. Lamellae up to 3 mm broad, greyish white when young, then gradually becoming pale brown, adnate, uneven and lighter at edge. Stipe 2–3.2 cm long, 3–4 mm thick, cylindrical, equal, solid, covered with smoky brown fibrous scales, slightly with white pruinose granules at apex, not enlarged at base. Basidiospores 8–9 × 5.5–6.5 μm, pale brown, with 7–10 small nodules. Pleurocystidia similar to cheilocystidia, 21–44 × 11–20 μm, thick-walled, especially thicker at apex.

In summer to autumn, solitary or scattered on very rotten wood of coniferous trees. Distributed in Northeast China.

This species is one of the few epixylous species within *Inocybe*, the macroscopic characteristics of this species are similar to those of *I. casimiri*, however the latter species has no pleurocystidia and has only thin-walled cheilocystidia.

1201 *Inocybe montana* Kobayasi

Pileus 0.4–0.6 cm broad, campanulate when young, hemispherical when mature, with un inconspicuously obtuse prominence at centre, pale grey brown to brown, subglabrous when young, squamulose when mature; squamules greyish white, appressed to slightly reflexed; margin slightly lacerate. Context white to greyish white, fleshy. Lamellae up to 1 mm broad, moderately close, adnate, greyish white when young, yellowish white when mature, concolourous at edge. Stipe 1.2–1.5 cm long, 1 mm thick, cylindrical, equal, pale brown when young, creamy yellow at apex, brown when mature, solid, white and slightly enlarged at base. Basidiospores 8–9 × 6–7 μm, with irregularly obtuse nodules, yellowish brown.

In summer, solitary on moss layer above ground in subalpine coniferous forests. Distributed in Qinghai-Tibet Regions, China.

This species has small fruiting bodies, small patchy squamules on pileus. The macroscopic characteristics are similar to those of *I. calospora*, however the basidiospores of the latter species are echinulate.

1202 *Inocybe nitidiuscula* (Britzelm.) Lapl.

Pileus 1.9–3 cm broad, conical when young, then campanulate to gradually plane, reflexed at margin with age, with a small umbo at centre, glabrous to fibrillose, dark brown at disc, gradually lighter to margin, finely rimulose, lacerate at margin with age. Context white or translucent, with a light earthy smell. Lamellae up to 3.5 mm broad, adnate, subdecurrent with age, moderately close, unequal, dirty white when young, brownish when mature, concolourous at edge. Stipe 3–6 cm long, 2–4 mm thick, cylindrical, equal, pinkish brown in upper portion, pale brown to greyish white in lower portion, enlarged and with white wooly mycelium at base. Basidiospores 9–11 × 5–6 μm, elliptical to nearly walnut-shaped, smooth, pale brown.

In summer to autumn, solitary or scattered on ground in broad-leaved forests. Distributed in Northeast China.

1203 *Inocybe olivaceonigra* (E. Horak) Garrido

Pileus 1.1–2.3 cm broad, conical when young, gradually plane when mature, with a blunt umbo at centre, often lacerate or reflexed at margin when mature, silky smooth, with a few appressed squamules, dark atrovirens when young, then olivaceous with dark grey tint, becoming lighter toward margin. Context with an earthy smell, fleshy, white, greenish near pileipellis. Lamellae 2.5–4.5 mm broad, adnate, close, milky white to white when young, then greyish white to brown, lighter at edge, uneven. Stipe 2.8–5 cm long, 1.5–3 mm thick, cylindrical, equal, solid, slightly enlarged at base, with white pruinose granules, longitudinally striate, light milky white to beige yellow, white at apex and base. Basidiospores 9–10 × 6–7.5 μm, obviously verruculose, brown.

In summer, scattered on ground under Fagaceae forests. Distributed in Central China.

1204 *Inocybe pachypleura* Takah. Kobay.

Pileus 0.9–1.1 cm broad, glabrous, campanulate to convex, ochraceous yellow, lighter at centre, inconspicuously rimulose, often finely lacerate at margin. Context 3–4 mm thick, with an earthy smell, white. Lamellae adnate, close, lighter at edge, white to greyish white. Stipe 2.8–3.5 cm long, 4–5 mm thick, cylindrical, equal, solid; enlarged and marginate at base up to 0.7 cm broad, with white pruinose granules, fleshy pink, white at base. Basidiospores 7–8 × 5.5–6.5 μm, weakly verruculose, brown.

In autumn, solitary on ground in broad-leaved forests. Distributed in Northeast China.

The distinguishing characteristics of this species are the glabrous pileus, fleshy pink colour and white pruinose granules on the stipe. It is macroscopically close to *I. grammata*, however the latter species has an obvious obtuse prominent at pileus centre.

1205 *Inocybe rimosa* (Bull.) P. Kumm.

Pileus 3–6.5 cm broad, campanulate when young, then plane, acutely prominent or umbonate at centre, straw yellow, finely ramulose to splitted. Context white to pale yellowish brown. Lamellae close, narrow, adnate to subfree, straw yellow, yellowish brown to olivaceous, lighter at edge. Stipe 6–9 cm long, 3–5 mm thick, cylindrical, equal, solid, white to yellow, furfuraceous at apex, gradually with fibrous squamules downwards, with fugacious velar remnants when young. Basidiospores 9.5–14.5 × 6–8.5 μm, long ellipsoid to phaseoliform, smooth, brown.

In summer to autumn, on ground in various broad-leaved and coniferous forests. Medicinal. Distributed in Northeast, Northern, Northwest, and Central China.

1206 *Inocybe terrigena* (Fr.) Kuyper

Pileus 1.6–2.4 cm broad, hemispherical at first when young, incurved at margin, with pale yellow cortinoid velar remnants, then campanulate, nearly plane to depressed at centre, yellow to brownish yellow, subglabrous, with appressed squamules radially spreading from centre to margin. Context fleshy, milky yellow, without conspicuous odour. Lamellae up to 5 mm broad, decurrent, olivaceous when young, yellowish brown when mature, close, subeven at edge. Stipe 3.5–5 cm long, 6–8 mm thick, cylindrical, equal, solid, glabrous at apex, with thick fibrillose squamules below annulus, yellow to yellowish brown. Basidiospores 8.5–9.5 × 4–5 μm, subphaseoliform, smooth, yellowish brown.

Solitary on roadside in coniferous or broad-leaved forests, uncommon. Distributed in Northeast and Northwest China.

This species is close to *I. paludinella*, however, the latter species has not a bulbous and integrally marginate stipe base.

1207 *Inocybe umbratica* Quél.

Pileus 1.8–2.2 cm broad, campanulate when young, gradually plane to revolute when mature, obviously prominent at centre, white to milky white, glabrous, dry; margin slightly exceeding lamellae, ramulose to serrate, without cortinoid velar remnants. Context thin, fleshy, white. Lamellae up to 2.5 mm broad, greyish white at first, then greyish brown, close, sinuate, thin, finely serrulate at edge, uneven, sometimes furcate. Stipe 5.5–6.5 cm long, 3–5 mm thick, cylindrical, solid, milky white, equal; base bulbous, integrally marginate, up to 6 mm thick at enlarged region; surface with minute white pruinose granules. Basidiospores 7–8 × 5–6 µm, oval to elliptical, verruculose, pale brown.

In summer to autumn, solitary on ground in broad-leaved forests. Distributed in Northeast China.

Agarics 4: Species with Names Beginning with K-L

1208 *Kuehneromyces lignicola* (Peck) Redhead

Pileus 1–3 cm broad, convex, gradually plane; margin incurved at first, then decurved, even to undulate, striate when wet; surface glabrous, hygrophanous, umbrinous, fading to yellowish brown toward margin. Context thin, yellowish brown, unchanging when injured. Lamellae adnate to sinuate, close, pale brown, becoming dark brown in later stage. Stipe 2–5 cm long, 2–3 mm thick, subequal, often curved at base, pale brown in upper portion, pruinose, glabrous to sparsely fibrillose in lower portion, dirty vinaceous brown to dark brown, interiorly full filled at first, then hollow. Partial veil cortinoid, fugacious, leaving some cortinoid residues on stipe. Basidiospores 5.5–8 × 3.5–4.5 μm, elliptical, smooth, hyaline.

In spring, fasciculate on rotten wood of conifers trees. Distributed in Northeast China.

1209 *Kuehneromyces mutabilis* (Schaeff.) Singer & A.H. Sm.

Pileus 2–6 cm broad, hemispherical to convex, gradually plane, often prominent at centre, incurved at margin; slightly sticky when wet, hygrophanous, glabrous or with inconspicuous white fibrils, yellow brown to umbrinous, often reddish-brown at disc, translucently striate at margin when wet. Context white to yellowish brown. Lamellae adnate or subdecurrent, lighter at first, then ferruginous brown. Stipe 4–10 cm long, 0.2–1 cm thick, central, cylindrical, equal, or tapering to base, whitish to yellowish brown and pruinose above annulus, dark brown and with recurved greyish white to brown squamules below annulus, without appendage or with white flocculent mycelium at base, becoming soft to hollow inside. Annulus superior, membranous. Basidiospores 5.5–7.5 × 3.5–4.5 μm, elliptical or oval, smooth, light ferruginous.

In summer to autumn, caespitose on fallen wood or stumps of broad-leaved trees. Edible; cultivatable. Distributed in Northeast, Northern, Northwest, Inner Mongolia, and Central China.

© Central China Farmer's Publishing House 2025
Y. Li et al., *Atlas of Chinese Macrofungal Resources*, https://doi.org/10.1007/978-981-96-0502-6_5

1210 *Laccaria alba* Zhu L. Yang & Lan Wang

Pileus 1–3.5 cm broad, white to dirty white, sometimes pink. Context thin, white. Lamellae pinkish. Stipe 3–5 cm long, 3–6 mm thick, subcylindrical, white to dirty white, glabrous to with small fibrillose squamules, with white mycelium at base. Basidiospores 7–9.5 x 7–9 µm, globose to subglobose, spinulose, with spines 1.5–2 µm long, hyaline.

In summer to autumn, on ground in forests. Edible. Distributed in Central China.

1211 *Laccaria ametystea* (Bull.) Murrill

Pileus 2–5 cm broad, flat spherical at first, then gradually plane, umbilicate at centre, bluish violaceous or lotus root pink to greyish violaceous, waxy, greyish white with violaceous tinge when dry, then undulate or valvate and strongly striate at margin, often squamulose, not viscid, radially sulcate. Context concolourous with pileus, thin. Lamellae adnate, slightly decurrent to subsinuate, wide, distant, unequal, concolourous with pileus or slightly darker, fading to yellowish brown when old. Stipe 3–8 cm long, 2–8 mm thick, subcylindrical, concolourous with pileus, tomentose, often curved at lower portion. Basidiospores 8.5–13 × 7–11.5 μm, oval to broadly elliptical, spinulate or verruculose, with spines 1.5–2.5 μm long, hyaline.

In summer to autumn, solitary or gregarious, and sometimes subcaespitose on ground in forests. Edible. Distributed widely in China.

1212 *Laccaria amethysteo-occidentalis* G.M. Muell.

Pileus 0.5–6 cm broad, hemispherical to gradually plane at first, often depressed at centre, violet when young, then gradually becoming paler to earthy yellow; firillose when fresh, rough when dried, incised at margin, estriate, hygrophanous. Context thin, violaceous when young, then gradually changing to earthy yellow, with a mild odour. Lamellae 0.5–1 mm broad, decurrent to free, deeply undulate to arched, unequal, distant, violet to greyish violaceous. Stipe 2–5 cm long, 2–5 mm thick, concolourous with pileus, fibrillose-squamulose, cylindrical, slightly thicker at base, fibrillose, longitudinally striate. Basidiospores 8.5–10.9 × 7.3–8.5 μm, wide elliptical, spinulose, with spines 0.5–1 μm long.

In summer to autumn, scattered on ground in broad-leaved forests and on sandy soil by roadside. Distributed in Northeast China.

1213 *Laccaria angustilamella* Zhu L. Yang & Lan Wang

Pileus 2–3 cm broad, plane, flesh-coloured to pale flesh-coloured, grey at centre, radially sulcate at margin. Context thin, flesh-coloured to white. Lamellae flesh-coloured, narrow. Stipe 4–6 cm long, 3–5 mm thick, subcylindrical, reddish brown, with small fibrillose squamules, with white mycelium at base. Basidiospores 8–11.5 × 8–11 μm, globose to subglobose, spinulose, with spines 2.5–5 μm long, subhyaline.

In summer to autumn, on ground in subtropical evergreen broad-leaved forests. Edible. Distributed in Central China.

1214 *Laccaria laccata* (Scop.) Cooke

Pileus 2.5–4.5 cm broad, thin, nearly oblate hemispherical, then gradually plane to reflexed, umbilicate at centre, fleshy pink, pale red brown or greyish bluish violaceous when fresh, hygrophanous when wet, and flesh-coloured to lotus-root-pink or light violaceous to eggshell colour when dried, glabrous to subglabrous, undulate or valvate and strongly striate at margin. Context concolourous with pileus or pinkish brown, thin. Lamellae adnate to subsinuate, distant, wide, unequal, flesh pink, pale red brown or greyish blue violaceous, white pruinose. Stipe 3.5–8.5 cm long, 3–8 mm thick, cylindrical, concolourous with pileus, subcylindrical or slightly flattened cylindrical, often curved at lower portion, solid, fibrous, tough, soft insides. Basidiospores 7.5–11 × 7–9 μm, subglobose, spinulose, hyaline or yellowish.

In summer to autumn, scattered or gregarious on ground and humus in coniferous forests and broad-leaved forests in middle and low altitudes, or on sandy slopes outside of the forests, sometimes subcaespitose. Edible. Distributed widely in China.

1215 *Laccaria longipes* G.M. Muell.

Pileus 0.7–3.2 cm broad, hemispherical, slightly depressed at centre, yellowish brown to orange red when fresh, earthy yellow when dried, glabrous, finely striate and incised at margin. Context thin, flesh pink to yellowish brown, with a mild odour. Lamellae 1.5–2 mm broad, decurrent, deeply undulate, thick, unequal, distant, flesh pink when fresh, reddish brown to dark grey when dried. Stipe 4–11 cm long, 1–3 mm thick, concolourous with pileus, cylindrical, slightly thicker at base, longitudinally fibrillose-striate, fibrillose-squamulose, with white mycelium at base. Basidiospores 7.3–8.7 μm, globose, densely spinulate, with spines 7.3–8.7 long.

In summer and autumn, scattered on ground in mixed forests with *Betula platyphylla* and *Larix gmelinii*. Distributed in Northeast.

1216 *Laccaria proxima* (Boud.) Pat.

Pileus 2–7 cm broad, flat hemispherical to nearly plane, depressed at centre, light latericeous, finely roughened to squamulose, hygrophanous when wet, subwavy and finely striate at margin. Context thin, light flesh pink. Lamellae adnate to decurrent, distant, wide, thick, pinkish, unequal. Stipe 8–12 cm long, 2–9 mm thick, subcylindrical, concolourous with pileus or brownish yellow, longitudinally fibrillose-striate, sericeous, often twisted, soft insides, lighter and with white tomentose mycelium at base. Basidiospores 7.5–9.5 × 6.5–8 μm, suboval or subglobose, spinulose, hyaline.

In summer to autumn, solitary or gregarious on ground in forests. Edible. Distributed in Northeast, Central, and Inner Mongolia China.

1218 *Lactarius akahatsu* Nobuj. Tanaka

Pileus 4–10 cm broad, flat hemispherical and slightly depressed at centre at first, then gradually plane, depressed to infundibuliform at centre, estriate at margin, pale orange, pale yellow, pale yellow brown, slightly viscid when wet, weakly zonate. Context pale yellow, with orange macules; odour mild. Lamellae adnate to short decurrent, subclose, orange, changing to green when injured, unequal. Latex orange. Stipe 3–4 cm long, 1–1.5 cm thick, fleshy, pale orange red, not scrobiculate, subequal. Basidiospores 8–9 × 6–7 µm, elliptical to subelliptical, subreticulate with connected ridges, amyloid, hyaline.

In summer, scattered to gregarious on ground in *Pinus* sp. forests. Edible. Distributed in Northeast, Central, and Southern China.

1217 *Laccaria vinaceoavellanea* Hongo

Pileus 2–5 cm broad, flat hemispherical to plane, often depressed at centre, fleshy brown, often squamulose, not viscid, long radially sulcate. Context thin. Lamellae adnate to subdecurrent, concolourous with pileus or darker. Stipe 4–8 cm long, 4–8 mm thick, subcylindrical, concolourous with pileus. Basidiospores 7.5–9 × 7.5–9 µm, globose to subglobose, spinulose, with spines 1.5–2.5 µm, subhyaline.

In summer to autumn, on ground in forests. Edible. Distributed widely in China.

1219 *Lactarius camphoratus* (Bull.) Fr.

Pileus 1–4 cm broad, convex, gradually becoming broadly convex or depressed at centre, often papillate, wet or dry, glabrous or pruinose, dark reddish brown, often fading to ferruginous brown or orange brown, crenulated at margin when old. Context light cinnamon to whitish, hard and brittle. Lamellae adnate or subdecurrent, close or clouded, whitish to light pink, often light red to cinnamon when mature. Latex milky white, whey-like. Stipe 1–5.5 cm long, 0.8–1 cm thick, equal, glabrous or fibrillose at base, concolourous with pileus or lighter. Basidiospores 7–8 × 6–7.5 µm, subglobose to wide elliptical, verruculose or with scattered ridges, not reticulate, hyaline to subhyaline, amyloid.

In spring to autumn, solitary, scattered or gregarious on ground in coniferous or broad-leaved forests. Medicinal. Distributed in Northeast, Northwest, Central, and Southern China.

1220 *Lactarius castaneus* W.F. Chiu

Pileus 4–9 cm broad, flat hemispherical to plane, greyish yellow, pale brown to brownish orange, glutinous, azonate. Context pale brown, bitter. Lamellae adnate to decurrent, close, white to pale yellow. Latex white, unchanging, bitter. Stipe 4–8 cm long, 0.5–1.5 cm thick, cylindrical or tapering upwards, pale yellow to whitish, glabrous, glutinous. Basidiospores 8.5–10.5 × 7–8.5 µm, wide elliptical, subhyaline, with amyloid ridges arranged in an incomplete reticulation.

In summer to autumn, on ground in coniferous forests. Distributed in Central, Qinghai-Tibet Regions, China.

1221 *Lactarius chichuensis* W. F. Chiu

Pileus 2–5 cm broad, plano-depressed, with or without spinous protrusions at centre, sometimes valved at margin, rugose, dry, glabrous, sometimes hygrophanous or white pruinose, radially translucently striate at margin with age, reddish brown, azonate. Context concolourous with pileus or lighter, or whitish, slightly bitter to bitter. Lamellae adnate to short decurrent, subclose to close, concolourous with pileus or darker. Latex white, then changing to clear water-like, without special odour. Stipe 2–5.5 cm long, 4–9 cm thick, central or eccentric, equal or tapering downwards. Basidiospores 6–7.5 × 5.5–7 µm, wide elliptical, with subparallel ridges and connected or isolated warts arranged in a zebra-pattern typically. Pleurocystidia 47–90 × 6–12 µm, rare to abundant.

In summer, scattered to gregarious on ground in Fagaceae forests. Distributed in Central China.

1222 *Lactarius cinnamomeus* W.E. Chiu

Pileus 3–6 cm broad, flat hemispherical to plane, greyish yellow, olive brown, pale yellow to cinnamon brown, glutinous when wet, azonate, radially wrinkles. Context dirty white, slightly bitter and peppery. Lamellae adnate to decurrent, close, white to beige with grey to orange tint. Latex white, unchanging, bitter and peppery. Stipe 2–5 cm long, 0.5–1 cm thick, cylindrical, concolourous with pileus. Basidiospores 6.5–8 × 5.5–6.5 µm, wide elliptical, subhyaline, with amyloid ridges and warts connected as an incomplete reticulation.

In summer to autumn, on ground in mixed forests. Distributed in Central China.

1223 *Lactarius deliciosus* (L.) Gray

Pileus 4–10 cm broad, flat hemispherical to plane, depressed at centre, subviscid when wet, yellowish brown to orange yellow, concentrically zonate, depressed at centre, incurved at margin. Context whitish to pale yellow or orange yellow, darker at stipe, becoming turquoise when injured, not peppery. Lamellae narrow, close, orange yellow, slowly changing to green when injured or aged. Latex scarce, orange to carrot coloured, unchanging, or becoming wine red when exposed to air, not peppery. Stipe 2–6 cm long, 0.8–2 cm thick, cylindrical, concolourous with pileus, with darker scrobicules. Basidiospores 7–9 × 5.5–7 μm, up to 12 × 9 μm when reticulation included, wide elliptical to oval, with an incomplete reticulation and discrete short ridges, subhyaline to yellowish, amyloid.

In summer to autumn, on ground in coniferous forests. Famous edible mushroom. Distributed widely in China.

1224 *Lactarius deterrimus* Gröger

Pileus 5–10 cm broad, orange red to orange yellow, partially with greenish tint, inconspicuously concentrically zonate. Context whitish, not peppery. Lamellae adnate, bright orange yellow, slowly changing to green when injured. Latex orange yellow to orange red, slowly changing to green after exuding from the wound. Stipe 3–6 cm long, 1–3 cm thick, cylindrical, lighter than pileus, subeven. Basidiospores 8–10 × 6–7 μm, wide elliptical to oval, subhyaline, incompletely reticulate and with discrete short ridges, amyloid.

In summer to autumn, on ground in forests of *Picea asperata*. Edible. Distributed in Northeast and Qinghai-Tibet Regions, China.

1225 *Lactarius glyciosmus* (Fr.) Fr.

Pileus 2–5 cm broad, plano-depressed, plane at margin, sometimes with spinous protrusions at centre, dry, subfelty, sometimes hygrophanous, sometimes zonate at margin, greyish white, greyish pink. Context whitish. Lamellae short decurrent, moderately close, light fleshy pink. Latex white. Stipe 2–5 cm long, 0.4–1 cm thick, central, equal or tapering upwards. Basidiospores 6–7.5 × 5.5–6.5 μm, wide elliptical, with subparallel ridges and connected or isolated warts arranged in a typical zebra-pattern. Pleurocystidia 47–90 × 6–12 μm, rare to abundant.

In summer, scattered to gregarious on ground in forests of *Betula*. Distributed in Northeast China.

1226 *Lactarius gracilis* Hongo

Pileus 1–3 cm broad, flat hemispherical to plane, brown, reddish brown to cinnamon, minutely cuspidate or acutely papillate at centre, obviously fringed with long hairs at margin. Context pale brown, not peppery. Lamellae with little latex, white. Stipe 4–5 cm long, 2–4 mm thick, cylindrical or gradually thicker downwards, concolourous with pileus or darker, with bristles at base. Basidiospores 7.5–8.5 × 6.5–7.5 μm, wide elliptical, subhyaline, with complete to incomplete reticulation, amyloid.

In summer to autumn, on ground in broad-leaved or mixed forests. Distributed widely in China.

1227 *Lactarius hatsudake* Nobuj. Tanaka

Pileus 3–6 cm broad, flat hemispherical to plane, greyish red to light red, unclearly zonate or azonate, depressed at centre, incurved at margin. Context light red, not peppery. Lamellae vinaceous red, slowly changing to bluish green when injured or mature. Latex little, vinaceous red, unchanging, not peppery. Stipe 2–6 cm long, 0.5–1 mm thick, slowly changing to bluish green when injured, not scrobiculate. Basidiospores 8–10 × 7–8.5 µm, wide elliptical, subhyaline, completely to incompletely reticulate, amyloid.

In summer to autumn, on ground in coniferous forests. Edible. Distributed widely in China.

1228 *Lactarius hirtipes* J.Z. Ying

Pileus 2–4 cm broad, flat hemispherical to plane, reddish brown to orange brown, depressed at centre, azonate. Context not peppery. Lamellae adnate to decurrent. Latex little, white, unchanging, slightly bitterish. Stipe 3–8 cm long, 3–6 mm thick, cylindrical or tapering upwards, concolourous with pileus or lighter, hirsute at base. Basidiospores 6.5–8 × 6–7.5 µm, subglobose to wide elliptical, subhyaline, completely to incompletely reticulate, amyloid.

In summer to autumn, on ground in broad-leaved forests. Distributed in Central China.

1229 *Lactarius imbricatus* M.X. Zhou & H.A. Wen

Pileus 2–5 cm broad, flat hemispherical to plane, grey to dirty grey, azonate, with blackish scales, depressed at centre, with a small pointed papilla. Context peppery. Lamellae with abundant latex, white, unchanging, peppery. Stipe 2–3 cm long, 0.3–0.6 mm thick, eccentric, cylindrical or tapering upwards, pale yellow to pale brown, glabrous. Basidiospores 7–9 × 5.5–6.5 µm, elliptical to wide elliptical, subhyaline, with ridges and warts arranged in a typical zebra-pattern, amyloid.

In summer to autumn, on ground in subalpine coniferous forests. Distributed in Qinghai-Tibet Regions, China.

1230 *Lactarius lignicola* W.F. Chiu

Pileus 3–5 cm broad, plane, then infundibuliform, brown to greyish brown, darker at centre, lighter at margin, obscurely zonate or azonate. Context white, peppery. Lamellae pale yellow to yellowish brown. Latex white, unchanging, bitter and peppery. Stipe 2–5 cm long, 0.5–1 cm thick, often eccentric, cylindrical, brown, glabrous. Basidiospores 7–8.5 × 6–7.5 μm, wide elliptical, subhyaline, with warts connected by fine lines, amyloid.

In summer to autumn, on rotten woods in broad-leaved or mixed forests. Distributed in Central China.

1231 *Lactarius lignyotus* Fr.

Pileus 4–10 cm broad, flat hemispherical at first, then gradually plane, brown to blackish brown, slightly depressed at centre, dry, subtomentulose, with blackish brown wrinkles. Context white, thick, slightly red when injured. Lamellae wide, distant, decurrent, unequal, white. Latex white to milky white, changing to light pink to light pinkish brown when exposed to air. Stipe 3–10 cm long, 0.4–1 cm thick, subcylindrical, concolourous with pileus, with blackish brown striae extending from decurrent lamellae, sometimes tomentose at base. Basidiospores 9–12.5 × 8.5–11 μm, globose to subglobose, with fine spines and ridged reticulation, hyaline, amyloid.

In summer to autumn, scattered on ground in forests, also reportedly on rotten woods of coniferous trees. Distributed in Northeast and Central China.

1232 *Lactarius necator* (Bull.) Pers.

Pileus 5–12.5 cm broad, hemispherical, incurved at margin at first, then gradually plane, sometimes depressed at centre, olivaceous brown or yellowish green, becoming darker to nearly black in later stage, viscid, tomentose. Context dirty white, gradually changing to brown, peppery. Lamellae adnate or decurrent, close, dirty white or olivaceous brown. Latex white. Stipe 3–7 cm long, 1–3 cm thick, concolourous with pileus. Basidiospores 7–8.5 × 6–7 μm, subglobose, ridged, hyaline, amyloid.

In summer to autumn, gregarious or solitary on ground in coniferous or broad-leaved forests. Useful as a spice. Distributed in Northeast and Central China.

1233 *Lactarius rufus* (Scop.) Fr.

Pileus 3–7 cm broad, plano-depressed, then subinfundibuliform shallowly, with or without spinous protrusions at centre, dry, often rugose, slightly felty, reddish brown, deep red, azonate. Context light red, concolourous with pileus or lighter. Lamellae decurrent, subclose, pale red brown. Latex white, unchanging, without special odour. Stipe 3–7 cm long, 0.5–1 cm thick, central, equal or tapering downwards. Basidiospores 6–7.5 × 5.5–6.5 μm, wide elliptical, with relatively complete reticulation composed of obvious ridges and warts and some isolated warts. Pleurocystidia 47–90 × 6–12 μm, rare to abundant.

In summer, scattered to gregarious on ground in forests of *Betula*. Distributed in Northeast China.

1234 *Lactarius torminosus* (Schaeff.) Gray

Pileus 5–10 cm broad, plano-depressed, plane at margin; slightly viscid when wet, with appressed long hairs, with some long hairs exceeding the margin, sometimes zonate, pink to pale red brown. Context whitish. Lamellae adnate to short decurrent, close, light pink. Latex white. Stipe 2–5 cm long, 1–2 cm thick, central, equal or tapering upwards, pink. Basidiospores 8–9.5 × 6–7 μm, with an incomplete reticulation composed of connected irregular ridges, and commonly with isolated warts. Pleurocystidia 60–70 × 7–10 μm, uncommon to common.

In summer to autumn, scattered to gregarious on ground in mixed forests. Distributed in Central China.

1235 *Lactifluus gerardii* (Peck) Kuntze

Pileus 2–10 cm broad, plane to revolute, often depressed and with spinous protrusions at centre, often radially rugose, subvelvety, dry, greyish yellow, yellowish brown, brown. Context 2–4 mm thick, white. Lamellae 0.7–1.2 cm broad, thick, very distant, decurrent, white. Latex white, unchanging, or changing to clear watery, mild. Stipe 3–8 cm long, 0.5–1.7 cm thick, often tapering downwards, concolourous with pileus or slightly darker. Basidiospores 8–11.5 × 7.5–10 μm, subglobose, rarely wide elliptical, with complete reticulation formed by relatively regular ridges, occasionally with isolated warts and free ridge terminals.

In summer to autumn, scattered on ground in broad-leaved forests. Edible. Distributed in Central and Southern China.

1236 *Lactifluus hygrophoroides* (Berk. & M.A. Curtis) Kuntze

Pileus 3–8 cm broad, depressed at centre, plane when mature, incurved at margin, obviously powdery-velvety, often irregularly rugose, orange brown, orange red, reddish brown. Context 3–5 mm thick, whitish, mild. Lamellae 2–10 mm broad, decurrent, distant, pale yellowish white, greyish yellow to greyish orange. Latex abundant, white, unchanging. Stipe 1–4 cm long, 0.6–1.5 cm thick, equal or tapering downwards, concolourous with pileus or lighter. Basidiospores 8–9.5 × 6.5–7.5 µm, elliptical, with an incomplete to subcomplete reticulation composed of connected ridges.

In summer to autumn, scattered on ground in mixed forests. Edible. Distributed in Central and Southern China.

1237 *Lactifluus piperatus* (L.) Roussel.

Pileus 5–13 cm broad, flat hemispherical and umbilicate at centre at first, finally depressed as infundibuliform, white or slightly tinted with dirty yellowish white or yellow, glabrous or even, not viscid or slightly viscid, brittle, azonate; margin incurved at first, then plane, gradually thinner and slightly reflexed outwards, sometimes undulate. Context thick, white, brittle, unchanging or slightly changing to light earthy yellow when injured, peppery. Lamellae white or eggshell-coloured, narrow, very close, unequal, furcate, subdecurrent, then changing to light earthy yellow. Latex white, unchanging. Stipe 3–6 cm long, 1.5–3 cm thick, stout, white, cylindrical, equal or tapering downwards, glabrous. Basidiospores 6.5–8.7 × 5.5–7 µm, subglobose or wide elliptical, verruculose or slightly rough, hyaline, amyloid.

In summer to autumn, scattered or gregarious on ground in coniferous and mixed forests. Edible. Distributed in Central and Southern China.

1238 *Lactifluus subpiperatus* (Hongo) Verbeken

Pileus 9–10 cm broad, shallowly infundibuliform, incurved at margin when young, slightly rugose, dry, white to milky white. Context 0.4 cm thick, white. Lamellae 3–4 mm broad, subdistant, decurrent. Latex white to creamy, abundant, peppery, slowly changing to yellow to brown, and staining the wound or context yellow to brown. Stipe 6.5–7 cm long, 1.5–2 cm thick, gradually thicker downwards, central to slightly eccentric, pruinose, concolourous with pileus. Basidiospores 5.5–7 × 5–6.5 µm, wide elliptical, elliptical, sparsely ridged and warted, subhyaline, amyloid. Pleurocystidia 45–55 × 5.5–6.3 µm, abundant, subcylindrical, clavate, obtuse at apex, with thick refractive contents.

Gregarious on ground in Fagaceae forests. Distributed in Central and Southern China.

1240 *Lactifluus tenuicystidiatus* (X.H. Wang & Verbeken) X.H. Wang

Pileus 6–10 cm broad, plane, depressed at centre, yellow to brownish yellow, tomentose. Context not peppery. Lamellae decurrent, distant, white to creamy. Latex abundant, white, mild to slightly peppery. Stipe 3–7 cm long, 1–3 cm thick, cylindrical or tapering upwards, white to beige, tomentose to subglabrous. Basidia 4-spored. Basidiospores 6.5–8 × 5.5–7 µm, elliptical, subhyaline, with an incomplete reticulation composed of ridges and warts, amyloid. Pleurocystidia thin-walled.

In summer to autumn, on ground in broad-leaved forests. Edible. Distributed in Central China.

1239 *Lactifluus subvellereus* (Peck) Nuytinck

Pileus 6–15 cm broad, flat hemispherical to hemispherical, depressed at centre, gradually plane, then shallowly infundibuliform, dry, white to dirty white, sometimes with faint earthy yellow macules, tomentulose, azonate, changing to cinnamon when dried or mature, incurved at margin at first, then plane. Context close, white, very peppery. Lamellae narrow, close, adnate to subdecurrent, white to pale yellow, cinnamon when injured or dried, unequal, often furcate. Latex white or slightly light creamy yellow, creamy yellow when dried, peppery. Stipe 3–8 cm long, 2–3.5 cm thick, generally stout, subcylindrical or slightly tapering downwards, white, short tomentose, cinnamon when dried. Basidiospores 7.5–10 × 5.5–8 µm, wide elliptical, oval to globose, with warts and connecting lines, hyaline, amyloid.

In autumn, solitary or gregarious on ground in broad-leaved or mixed forests. Distributed in Northeast and Central China.

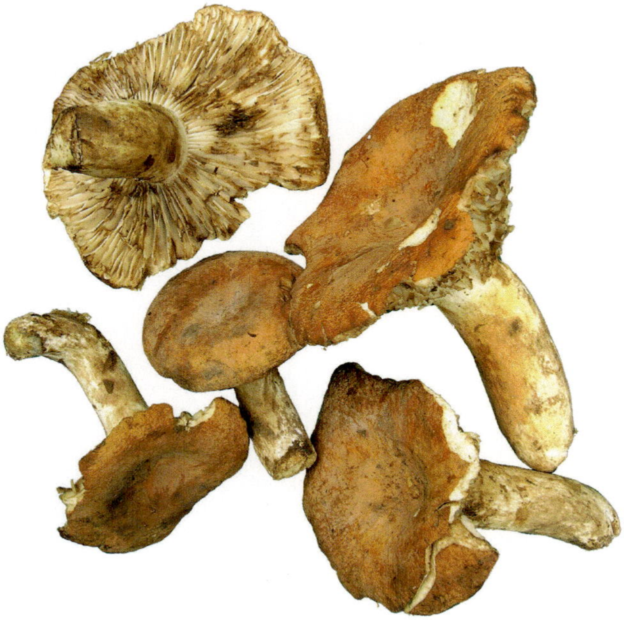

1241 *Lactifluus volemus* (Fr.) Kuntze

Pileus 4–11 cm broad, flat hemispherical at first, then gradually plane to umbilicate at centre, expanded to subinfundibuliform, orange red, reddish brown, chestnut brown, yellowish brown, amber brown, deep birchleaf-pear coloured or dark latericeous, mostly pruinose, not viscid, or slightly viscid when wet, azonate, glabrous or slightly tomentulose, incurved at margin at first, then plane. Context milky white, changing to pale brown when injured, hard and brittle, thick and compacted, not peppery. Lamellae white or pale yellow, changing to brownish yellow when injured, close, adnate to subdecurrent, furcate at stipe, unequal, exuding a large amount of white latex when injured. Latex white, unchanging. Stipe 3–10 cm long, 1–2.5 cm thick, subcylindrical or slightly tapering downwards, concolourous with pileus or lighter, subglabrous or tomentulose. Basidiospores 8.5–11 × 8–10 µm, subglobose or globose, reticulate and finely verruculose, hyaline to pale yellow, amyloid.

In summer to autumn, scattered, gregarious to rarely solitary on ground in *Pinus* forests or mixed forests, often forming mycorrhiza with *Pinus*. Edible. Distributed widely in China.

1242 *Lactocollybia epia* (Berk. & Broome) Pegler

Pileus 1–3 cm broad, flat hemispherical to plane, glabrous, white to creamy. Context white. Lamellae adnate to decurrent, white to cream, close. Stipe 1–6 cm long, 2–5 mm thick, concolourous with pileus, often eccentric. Basidiospores 6.5–8.5 × 4–5.5 µm, subamygdaliform, smooth. Pleurocystidia 30–65 × 8–18 µm, subfusiform, abundant.

In summer to autumn, on rotten wood in tropical and south subtropical forests. Distributed in Southern China.

1243 *Lentinellus auricula* (Fr.) E. Ludw.

Pileus 3–4 cm broad, auricular, conchate or pleurotoid, tomentose at base, white, creamy white or dirty white, yellow when dried; margin incurved and lacerate, or forming pileoli for strongly incurving, estriate. Context thin, white, yellow when dried. Lamellae up to 2.5 mm broad, adnate or decurrent to base, close to very-close, white to dirty white, finely serrulate at edge, yellow to yellow brown when dried or aged. Stipe absent, with a common attachment point from which the superimposed fruiting bodies growing. Taste weakly peppery or sour. Basidiospores small, 2–3 × 2.5–3.5 μm, subglobose to globose, scatteredly finely warted, amyloid.

In summer, superposed on rotten woods of evergreen broad-leaved forests. Distributed in Central China.

1244 *Lentinellus bambusinus* T.H. Li et al.

Pileus 2–4.7 cm broad, yellowish white to pale yellow, leathery, concave above stipe to form a lateral infundibuliform shape, tomentose. Context thin, unchanging when injured. Lamellae long decurrent, unequal, close, white to pale yellow. Stipe 3–5 cm long, 2–4.5 mm thick, eccentric to sublateral, white to pale yellow, tomentose, longitudinally striate, leathery, solid. Basidiospores 5.8–7.2 × 2.8–3.6 μm, elliptical, hyaline, nearly smooth to slightly asperulate, amyloid.

In summer to autumn, on bamboo stems. Distributed in Central China.

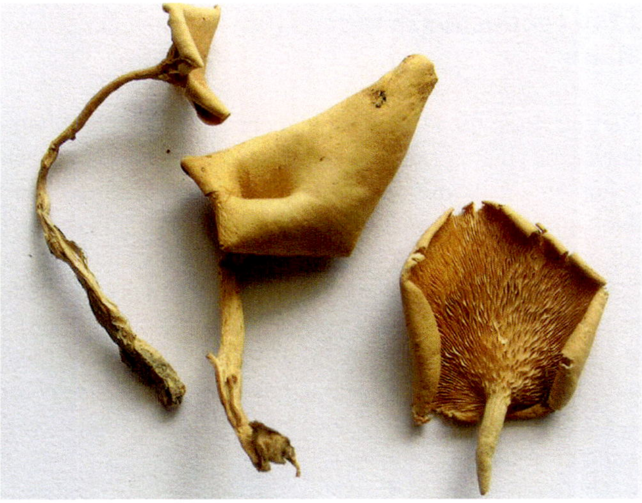

1245 *Lentinellus brunnescens* Lj.N. Vassiljeva

Pileus 3–5 cm broad, plane to convex, pale brown to pale yellowish brown, hirsute-tomentose all over, densely tomentose at base, gradually darker to yellowish brown to brown, estriate and incurved at margin. Context white, slightly yellowish or brownish, thick, especially tougher and more tenacious at base. Lamellae adnate to decurrent, close to very close, thin, yellow brown, darker when dried, neatly serrate at edge. Stipe lateral, 1–2 cm long, 1–1.5 mm thick, stout, concolourous with pileus or slightly darker to yellow brown or brown, densely hirsute. Taste acrid. Basidiospores 4.5–5 × 4.5–5 μm, globose to subglobose, hyaline, strongly amyloid.

Solitary or gregarious on rotten woods of broad-leaved trees in mixed forests. Distributed in Northeast China.

1246 *Lentinellus castoreus* (Fr.) Kühner & Maire

Pileus 2–5 cm broad, pleurotoid, ochraceous brown, salmon brown, or slightly with pinkish brown tint, incurved when young, gradually more tomentose toward base, densely and thick tomentum at base, or densely felty, dirty white or greyish white or brownish. Context thin, dirty white, thick. Lamellae deeply decurrent, close, flesh coloured to pale brown; entire at edge when young, gradually undulate. Stipe abscent, with a wide base, with pale reddish brown to brown tomentum. Odour weak; taste slightly peppery. Basidiospores 4–5 × 3–3.5 μm, elliptical to wide elliptical, hyaline, thin-walled, verruculose, amyloid.

On rotten woods in mixed forests. Distributed in Qinghai-Tibet Regions, China.

1247 *Lentinellus cochleatus* (Pers.) P. Karst.

Pileus 3–6 cm broad, spatulate or plane at first, then cordiform or infundibuliform, glabrous, pale yellow brown or umbrinous, slightly incurved at margin. Context white or slightly with brownish tint, nearly corky-leathery or leathery. Lamellae decurrent, close, pale yellow brown to cinnamon, serrate at edge. Stipe 3–8 cm long, 0.3–1.2 cm thick, lateral or eccentric, concolourous with pileus or slightly lighter, tough, spirally twisted and confluent with other stipes at base, with long decurrent lamellae becoming deep furrows or ridges downwards. Basidiospores small, 3–4 × 4–5 μm, wide elliptical to elliptical, verruculose, hyaline, amyloid.

On rotten woods in mixed or broad-leaved forests. Distributed in Northeast, Northwest, and Central China.

This species is similar to *L. vulpinus*, however the fruiting bodies of *L. vulpinus* are usually lignified, hard when dried, while *L. cochleatus* has a longer stipe, which is non-lignified, brittle when dried.

1248 *Lentinellus flabelliformis* (Bolton) S. Ito

Pileus 0.8–2.5 cm broad, glabrous, plane, pale brown, cinnamon brown to hazel; margin entire when young, serrate or incised with age when exposing the wide and distant lamellae, hygrophanous. Context white or pale brown, thin.

Lamellae subdecurrent or decurrent, distant, serrate at edge, pale brown when fresh. Stipe 0.1–1 cm long, 2–6 mm thick, glabrous, pale brown, brown, dark reddish brown or concolourous with pileus. Odour weak; taste slightly peppery. Basidiospores 4.5–6.5 × 3.5–5 μm, wide elliptical, verruculose scatteredly, hyaline, amyloid.

Gregarious on withered barks to trunk wood in mixed forests. Distributed in Northeast, Northwest, and Central China.

1249 *Lentinellus jilinensis* Yu Liu & T. Bau

Pileus 2–3 cm broad, flabelliform or subspatulate or nearly plane when young, concave to umbilicate at centre, glabrous, cinnamon to light pinkish brown, reddish brown, dark reddish brown to blackish brown when dried, darker and slightly incurved at margin. Context thin, soft, whitish brown to pale yellow brown. Lamellae 2–2.5 mm thick, decurrent, distant, light pinkish brown to cinnamon, finely serrulate at edge. Stipe 2–4 cm long, 3–7 mm thick, shallowly longitudinally furrowed, eccentric to lateral, cinnamon to light pinkish brown. Tast acrid. Basidiospores 4–5 × 3.5–4.5 μm, elliptical to oval, verruculose, hyaline, amyloid.

In Autumn, solitary or gregarious in small groups on rotten woods of conifers or deciduous trees. Distributed in Northeast China.

This species is similar to *L. cochleatus*, however the fruiting bodies of this species have a segregate stipe rather than a stipe confluent with other stipes, and has no obviously deep longitudinal furrows or ridges on stipe.

1250 *Lentinellus sinensis* R.H. Petersen

Pileus 3–10 cm broad, semicircular to broadly cuneate, convex to conchate, concave or sulcate near stipe, glabrous, cinnamon to yellow brown when fresh, red brown to tobacco brown when dried, plane to slightly incurved at margin. Context often hygrophanous, white to pale brown, fibrous. Lamellae decurrent, very close, narrow; nearly entire at margin, sometimes undulate or inconspicuously serrulate. Stipe 4–10 mm long, 4–8 mm thick, cylindrical, concolourous with pileus in upper portion, gradually darker downwards to dark brown at base. Taste weakly peppery. Basidiospores 2.5–3.5 × 2–2.5 μm, wide elliptical to subglobose, verruculose, hyaline, strongly amyloid.

On rotten wood of deciduous trees in mixed or broad-leaved forests. Distributed in Northeast China.

1251 *Lentinellus ursinus* (Fr.) Kühner

Pileus 3–10 cm broad, pleurotoid, dry, pale brown, cinnamon or reddish brown, with short finely spinulose firils from middle to base, more densely so at base, hazel to brown, pale brown or cinnamon at margin, slightly hygrophanous; margin incurved, entire or serrated. Context pale brown, soft. Lamellae 4–5 mm broad, adnate to decurrent, close, pale brown to cinnamon brown, undulate or serrate at edge. Stipe abscent. Attachment base wide, usually with greyish brown thick bristles. Taste strongly sour or peppery. Basidiospores small, 3.5–4.5 × 2.8–3.2 μm, wide elliptical to elliptical, verruculose, hyaline, amyloid.

In summer to autumn, on rotten woods of broad-leaved trees such as *Acer* and *Betula*. Distributed in Northeast, Northern, Central, Northwest, and Qinghai-Tibet Regions, China.

1252 *Lentinellus vulpinus* (Sowerby) Kuhner & Maire

Pileus 2–4 cm broad, pleurotoid, pale brown to pale yellowish brown, reddish brown when mature or dried, with dense and thick pale yellow to pale brown tomentum when young, incurved at margin, radially sulcate or striate. Context 1–6 mm thick, dirty white or pale brown, nearly corky. Taste acrid or weakly sour. Lamellae thin, decurrent, close, pale yellow brown to pale red brown; slightly serrate at edge. Stipe 1.5–2.5 cm long, 0.8–1.5 cm thick, lateral, usually confluent with several other stipes from different fruiting bodies, with pale brown longitudinal striae on confluent stipes, lignified, longitudinally furrowed from the decurrent lamellae on lower portion of stipe. Basidiospores 3.5–4 × 3–3.5 μm, elliptical, verruculose, hyaline, amyloid.

Caespitose on rotten woods of broad-leaved trees such as *Ulmus pumila* in mixed forests. Distributed in Northeast China.

1253 *Lentinula edodes* (Berk.) Pegler

Pileus 5–12 cm broad, flat hemispherical to plane, pale brown, dark brown to dark cinnamon, with darker scales at disc and lighter or dirty white scales at margin, tomentose or flocculent, cracked or areolate in dry conditions; margin incurved at first, then plane, with pale brown flocculent partial veil connecting margin to stipe in early stage, appendiculate with partial velar remnants at margin when pileus expanded. Context thick, white, soft and tough. Lamellae white, close, sinuate, unequal. Stipe 3–10 cm long, 0.5–3 cm thick, central or eccentric, often curved to one side, solid, tough, fibrous. Annulus narrow, fugacious. Stipe surface with fibrillose scales below annulus. Basidiospores 4.5–7 × 3–4 μm, elliptical to oval, smooth, hyaline.

In autumn, scattered or solitary on fallen wood of broad-leaved trees. Famous edible fungus. Distributed in Northeast, Central, and Southern China.

1254 *Lentinus sajor-caju* (Fr.) Fr.

Pileus 4–17 cm broad, soft leathery, hardened when dried, convex-depressed to cyathiform, infundibuliform or subflabelliform, dirty white at first, often with greyish macules, then pale yellow, greyish yellow brown to greyish brown, glabrous to tomentose, sometimes with dark appressed squamules, often striate, cracked with age. Context tough, hard when dried, corneous when dried, white. Lamellae long decurrent, dirty white, concolourous with pileus to darker, thin, narrow, very close. Stipe 1.5–5 cm long, 0.8–1.5 cm thick, central, eccentric or lateral, cylindrical, hard, solid, white to concolourous with pileus. Annulus obvious to unobvious, white to yellowish brown when young, thick, often with annular marks when annulus falling off in later stage. Basidiospores 5–8 × 2–2.5 μm, cylindrical, often curved, hyaline, inamyloid.

In summer to autumn, solitary to gregarious on broadleaved trees and rotten woods in mixed forests. Distributed in Southern China.

In China, it was once to use the name *L. sajor-caju* mistakenly for a cultivated edible mushroom *Pleurotus pulmonarius*, however the latter species belongs to *Pleurotus* genus and has not any annulus on stipe.

1255 *Lentinus squarrosulus* Mont.

Pileus 4–13 cm broad, thin and flexible, convex-depressed to deeply infundibuliform, greyish white, pale yellow or slightly brownish, dry, squamules floccose, concentrically arranged; reflexed to appressed, grey to brown, evanescent with age; margin incurved at first, thin, then lobed or lacerate. Context thick, leathery, white. Lamellae decurrent, furcate, sometimes slightly intertwined near stipe, white to pale yellow, close, thin. Stipe 1–3.5 cm long, 0.4–1 cm thick, cylindrical, subcentral to eccentric or sublateral, often tapering downwards, solid, concolourous with pileus, often slightly darker at base, with floccose squamules. Basidiospores 5.5–8 × 1.7–2.5 µm, long ellipsoid to subrectangular, smooth, hyaline, inamyloid.

Gregarious, caespitose or subsuperposed on rotten wood in mixed or broad-leaved forests. Edible when young. Distributed in Southern China.

1256 *Lentinus suavissimus* Fr.

Pileus 7–12 cm broad, broadly convex-depressed to deeply umbilicate or subinfundibuliform, pale yellow to orange yellow, yellowish brown, glabrous; margin incurved, undulate or valvately lacerate. Context thick, white to pale yellow, tough. Lamellae often decurrent, often interlaced to reticulate at stipe apex, white or whitish, yellowish brown when dried, narrow, close, serrulate at edge. Stipe 0.5–3 cm long, 0.5–1.3 cm thick, central, eccentric, lateral or abscent, cylindrical, slightly enlarged at base, solid, concolourous with pileus to dark reddish brown, glabrous, with a reticulation extending from decurrent lamellae in upper portion, sometimes long flocculent at base, with a base sometimes growing into the bark of substrate. Basidiospores 6–8 × 2.5–3.4 µm, subcylindrical, smooth, hyaline, inamyloid.

In summer to autumn, on dead wood of broad-leaved trees. Edible when young. Distributed in Northeast, Northwest, and Central China.

1257 *Lepiota aspera* (Pers.) Quél.

Pileus 4–10 cm broad, nearly plane, dirty white to yellow brown, with pyramidal or granular dark scales. Context white, fleshy. Lamellae free, dirty white, close, unequal. Stipe 5–12 cm long, 0.5–2 cm thick, dirty white, subglabrous above annulus, with pale brown, conical, fugacious scales below annulus. Annulus superior, membranous. Basidiospores 5.5–7.5 × 2–3 µm, elliptical to subcylindrical in side view, subcylindrical in frontal view, smooth, hyaline, dextrinoid. Pileal scales consisting of inflated elements joining into rosary arrangement.

In summer to autumn, in parks or woods. Distributed widely in China.

1258 *Lepiota brunneoincarnata* Chodat & C. Martín

Pileus 2.5–6 cm broad, subconical or campanulate, then plane, prominent at centre, pinkish flesh to pinkish brown, with pinkish brown, pinkish brown or dark purplish brown scales; brown at disc, white or dirty white at marginal areas without scales. Context thin, white. Lamellae free, white to milky white, close, unequal. Stipe 3–6 cm long, 4–8 mm thick, subcylindrical, hollow, concolourous with pileus, often with scales below annulus, lighter above annulus, whitish at apex. Annulus superior, often leaving membranous veler vestiges only. Basidiospores 6.5–9 × 4–5 μm, oval to wide elliptical, smooth, hyaline, dextrinoid.

In summer to autumn, on ground in broad-leaved forests. Extremely poisonous, containing phallotoxin and amatoxin. Distributed in Northeast, Northwest, and Southern China.

1259 *Lepiota castanea* Quél.

Pileus 2–4 cm broad, subcampanulate to flat at first, then plano-depressed, umbonate at centre, drab to light chestnut brown, darker at centre, with granular squamules all over. Context thin, dirty white. Lamellae free, yellowish white, unequal, close. Stipe 2–4 cm long, 2–4 mm thick, thin, cylindrical, hollow; subglabrous above annulus, dirty white, concolourous with pileus below annulus, with small brown squamules in annularly zoned arrangement. Annulus inconspicuous, superior. Basidiospores 9–12.5 × 4–5.5 μm, subfusiform, smooth, hyaline, dextrinoid.

In summer, on ground in coniferous forests. Distributed in Northeast and Central China.

1260 *Lepiota clypeolaria* (Bull.) P. Kumm.

Pileus 3–9 cm broad, dirty white, with pale yellow, yellowish brown, pale brown to umbrinous scales. Context thin, fleshy, white. Lamellae white. Stipe 5–12 cm long, 0.4–1 cm thick, subglabrous and white above annulus, densely with white to pale brown tomentose scales below annulus, often with white rhizomorphs at base. Annulus white, floccose to submembranous, fugacious. Basidiospores 11–15 × 4.5–7 μm, fusiform or subamygdaliform in side view, smooth, hyaline.

In summer to autumn, on ground in forests. Distributed widely in China.

1261 *Lepiota coloratipes* Vizzini et al.

Pileus 0.8–3 cm broad, dirty white to beige to pale brown, even when mature. Context thin, fleshy, white. Lamellae white to beige. Stipe 1.5–4 cm long, 1–3 mm thick, light flesh coloured to pale brown, sometimes dirty white. Annulus superior, cortinoid, fugacious. Basidiospores 3–4 × 2.5–3 μm, wide elliptical to elliptical, smooth, hyaline, dextrinoid. Pileus epicutis composed of elements in hymeniform arrangement.

In summer to autumn, on humus in broad-leaved or mixed forests. Distributed in Central China.

1262 *Lepiota cristata* (Bolton) P. Kumm.

Pileus 1–7 cm broad, white to dirty white, with reddish brown to brown scales, with an obtuse reddish brown smooth prominence at centre. Context thin, white, with a disgusting odour. Lamellae free, white. Stipe 1.5–8 cm long, 0.3–1 mm thick, white, then changing to reddish brown. Annulus superior, white, fugacious. Basidiospores 5.5–8 × 2.5–4 μm, ergot-shaped or subtriangular in side view, hyaline, dextrinoid. Pileus scales composed of cells in hymeniform arrangement.

Solitary or gregarious on ground in forests, roadsides, and lawn. Poisonous. Distributed widely in China.

1263 *Lepiota cristata* var. *macrospora* (Zhu L. Yang) J.F. Liang & Zhu L. Yang

Pileus 1.5–5 cm broad, white to dirty white, with reddish brown to brown scales. Context dirty white to pale red brown, often with a disgusting odour. Lamellae free, white. Stipe 2.5–7 cm long, 2–4 mm thick, reddish brown in lower portion, lighter in upper portion, hollow. Annulus superior, white, membranous, fugacious. Basidiospores 6–8 × 3–4 µm, subtriangular in side view, with a dorsal spur at base, ventricose in frontal view, obviously inflated on both sides, hyaline, dextrinoid.

Solitary or gregarious on ground in forests, roadside, and lawn. Poisonous. Distributed widely in China.

1264 *Lepiota cristatanea* J.F. Liang & Zhu L. Yang

Pileus 1.5–4.5 cm broad, hemispherical to plane and prominent at centre, white to dirty white, with reddish brown, brown to purplish brown patchy scales, with an obtuse reddish brown to dark brown umbo at centre. Context thin, white, often with a disgusting odour. Lamellae free, white. Stipe 2–5.5 cm long, 2–7 mm thick, subcylindrical, pale red brown to yellowish brown. Annulus superior, white, membranous, fugacious. Basidiospores 4–5.5 × 2.5–3 µm, subtriangular in side view, nearly truncated at base, with an inconspicuous spur, ventricose abdominally, oval in frontal view, hyaline, dextrinoid.

Solitary or gregarious on ground in forests, roadside, and lawn. Poisonous. Distributed in Central China.

1265 *Lepiota erminea* (Fr.) P. Kumm.

Pileus 3–5 cm broad, hemispherical at first, then gradually plane, obviously umbonate at centre, white, with yellowish brown scales, sulcate at margin. Context thin, fleshy, white. Lamellae white, close, free. Stipe 6–9 cm long, 3–6 mm thick, cylindrical, gradually thicker to base, white. Annulus superior, white, fugacious. Basidiospores 12.5–15.5 × 4.5–5.5 µm, subfusiform, smooth, hyaline, dextrinoid.

In summer to autumn, on ground in forests. Distributed in Northeast, Northern, and Central China.

1266 *Lepiota fuscovinacea* F.H. Møller & J.E. Lange

Pileus 3–5 cm broad, densely with pale brown appressed or slightly reflexed tomentose to flocculent scales; margin often incurved, appendiculate with flocculence. Context white, unchanging when injured. Lamellae free, white to beige. Stipe 4–6 cm long, 3–7 mm thick, dirty white, pink to fleshy pink, with brown appressed or slightly revolute tomentose to flocculent scales. Annulus fugacious, easily broken as veler remnants hanging at pileus margin. Basidiospores 4.5–6 × 2–2.5 µm, elliptical in side view, smooth, hyaline, dextrinoid.

In summer to autumn, solitary or gregarious on ground in forests or roadside. Distributed widely in China, but relatively rare.

1267 *Lepiota helveola* Bres.

Pileus 1–4 cm broad, flat hemispherical at first, then plane, slightly prominent at centre, densely with reddish brown or brown squamules often arranged in concentrical zones, with denser scales at centre. Context white. Lamellae free, close, white. Stipe 2–6 cm long, 3–5 mm thick, cylindrical, pale yellowish brown, slightly enlarged at base. Annulus superior, small and fugacious. Basidiospores 5–9 × 3.5–5 µm, elliptical, smooth, hyaline, dextrinoid.

In spring to autumn, solitary or gregarious in forests and grasslands at forest edges. Poisonous. Distributed in Northern, Northwest, and Central China.

1268 *Lepiota nivalis* W.F. Chiu

Pileus 1.2–2.5 cm broad, flat hemispherical to plane, white, with radially fibrillose scales. Context thin, fleshy, white. Lamellae free, white. Stipe 2–3 cm long, 2–3 mm thick, subcylindrical, white. Annulus white, membranous. Basidiospores 6–7.5 × 3.5–4.5 μm, amygdaliform in side view, oval in frontal view, smooth, hyaline, dextrinoid. Pileus epicutis composed of repent radially arranged hyphae in 3–5 μm diam.

On ground in forests. Distributed in Central China.

1269 *Lepiota pseudogranulosa* Velen.

Pileus 1.5–2.5 cm broad, hemispherical at first, then campanulate to flat hemispherical, with pale yellowish brown or light pink yellow and fugacious granular scales, white, with flocculent velar remnants at margin. Context white. Lamellae free, moderately close, white. Stipe 3–7 cm long, 1.5–2.5 mm thick, subcylindrical, with powdery-granular scales. Annulus superior, fugacious. Basidiospores 4.5–5.5 × 2–3 μm, elliptical, smooth, hyaline, dextrinoid.

In summer to autumn, scattered or solitary on ground in forests. Edible. Distributed in Southern China.

1270 *Lepiota rubrotincta* Peck

Pileus 3–6 cm broad, reddish brown, brick red to dark reddish brown, radially lacerate or cracked to form scales. Context thin, membranous to fleshy, white. Lamellae free, white. Stipe 3.5–8 cm long, 0.5–1 cm thick, subcylindrical, glabrous. Annulus superior, membranous, white. Basidiospores 7.5–9.5 × 4.5–5 μm, amygdaliform in side view, suboval in frontal view, smooth, hyaline, dextrinoid. Terminal elements of pileus scales lanceolate to subcylindrical.

On lawn or ground in forests. Edibility unknown. Distributed widely in China.

1271 *Lepiota shixingensis* Z.S. Bi & T.H. Li

Pileus 3.5–4 cm broad, suboval, campanulate, convex with a central prominence to plane with a central prominence, orange to orange brown, dark brown at centre, squamulose and hirsute, usually exuding many yellow, orange to purplish brown droplets on surface. Context white, unchanging when injured. Lamellae free, moderately close, unequal, yellow. Stipe 4–7 cm long, 5–7 mm thick, cylindrical, white in upper portion, pale brown to concolourous with pileus in lower portion, cylindrical, hollow, tomentose and hirsute. Annulus superior, white. Basidiospores 5–7 × 3.5–4.5 μm, subelliptical, with a germ-pore, hyaline, smooth, dextrinoid.

Scattered on ground containing plant residues in forests or on rotten wood. Distributed in Southern China.

1272 *Lepiota squamulosa* T. Bau & Yu Li

Pileus 0.5–2 cm broad, hemispherical, then plane, depressed at centre, densely with pink scales, with universal velar remnants at margin. Context white, thin. Lamellae white, moderately close, sinuate. Stipe 1.8–3.5 cm long, 1–2 mm thick, central, fibrous, densely with scales similar to those on pileus surface on middle and lower portions, white at apex. Annulus superior, fugacious. Basidiospores 4.5–5.5 × 2.5–3.5 μm, elliptical, smooth, hyaline, inamyloid.

In summer, on debris or leaf litter in forests. Distributed in Northeast, Inner Mongolia, and Northern China.

1273 *Lepista caespitosa* (Bres.) Singer

Pileus 3–8 cm broad, convex, then nearly plane, white, subundulate at margin. Context white. Lamellae adnate to subdecurrent, free in later stage, close, white to pinkish. Stipe 2–5 cm long, 0.8–1 cm thick, enlarged at base. Basidiospores 5.5–6.5 × 3.5–4.5 μm, oval to elliptical, warty, hyaline.

In summer to autumn, gregarious to caespitose on ground or humus in broad-leaved forests and mixed forests. Edible. Distributed in Northern, Northeast and Northwest China.

1274 *Lepista flaccida* (Sowerby) Pat.

= *Clitocybe inversa* (Scop.) Quél.

Pileus 2.5–8 cm broad, convex, then gradually plane, depressed at centre to shallowly infundibuliform, incurved at margin at first, sometimes undulate to slightly lacerate at margin, moist, glabrous, hygrophanous, pale orange brown, pinkish brown to cinnamon or red brown, lighter toward margin. Context thin, concolourous with pileus, fragrant, taste mild. Lamellae decurrent, moderately wide, clouded to close, light apricot yellow. Stipe 2–7 cm long, 5–7 mm thick, equal or slightly enlarged at base, moist, finely striate, concolourous with pileus, with pale yellow mycelium at base. Basidiospores 4–4.5 × 3.5–4 μm, subglobose to wide elliptical, spinulose, creamy yellow.

In autumn to early winter, scattered or gregarious on ground in coniferous and broad-leaved forests. Edible. Distributed in Northeast, Northwest, and Qinghai-Tibet Regions, China.

1275 *Lepista irina* (Fr.) H.E. Bigelow

Pileus 4–13 cm broad, flat hemispherical to plane, slightly prominent at centre, white, cream or light flesh-coloured to beige, glabrous, dry, pale yellow to pale brown at disc, incurved at margin. Context thick, soft, dirty white to flesh coloured. Lamellae white, dirty white to flesh, close to moderately close, adnate to slightly sinuate, unequal. Stipe 5–12 cm long, 1–2 cm thick, cylindrical, dirty white to weakly fleshy pink, longitudinally sulcate and fibrillose-squamulose, solid. Basidiospores 7.5–9.5 × 4–5.5 μm, elliptical to wide elliptical, nearly smooth or asperulate, hyaline.

In summer to autumn, on ground in coniferous forests, broad-leaved forests or mixed forests. Edible. Distributed in Northeast, Inner Mongolia, Northwest, and Qinghai-Tibet Regions China.

1276 *Lepista nuda* (Bull.) Cooke

Pileus 3–12 cm broad, flat hemispherical to plane, sometimes depressed at centre, moist, glabrous, initially bluish violaceous to lilac violaceous, then brownish violaceous, incurved at margin. Context thick, soft, lavender, white when dried. Lamellae adnate to subdecurrent, unequal, close, bluish violaceous or concolourous with pileus. Stipe 4–8 cm long, 0.7–2 cm thick, cylindrical, slightly enlarged at base, blueish violaceous or concolourous with pileus, glabrous or longitudinally striate on lower portion, slightly elastic, solid. Basidiospores 5–8 × 3–5 μm, nearly smooth or asperulate, hyaline.

In autumn, gregarious, subcaespitose, and scattered on ground in mixed forests. Edible. Distributed in Northeast, Northwest, Central, and Inner Mongolia China.

1277 *Lepista personata* (Fr.) Cooke

Pileus 5–16 cm broad, flat hemispherical or convex at first, then gradually plane, cream to light purplish pink, gradually fading to dirty white. Context thick, white to greyish white. Lamellae free to sinuate, close, pinkish, cream to pale brown. Stipe 5–7 cm long, 1.5–3 cm thick, usually bulbous at base, rarely tapering to base, covered with pale lavender fibrillose scales. Basidiospores 7.5–8.5 × 4–5 μm, elliptical, asperulate, hyaline.

In summer to autumn, gregarious on grassland or on edge of forests. Edible and medicinal. Distributed in Northeast, Inner Mongolia and Northwest China.

1278 *Lepista sordida* (Schumach.) Singer

Pileus 4–8 cm broad, hemispherical when young, then plane, sometimes depressed at centre, translucent or hygrophanous when wet, violet when fresh, gradually fading to yellow brown after dehydration; margin incurved, inconspicuously striate, often undulate or valvate. Context light violet, thin, hygrophanous. Lamellae adnate, sometimes subsinuate or subdecurrent, moderately close, light violaceous. Stipe 4–6.5 cm long, 0.3–1.2 cm thick, violet, solid, curved at base. Basidiospores 7–9.5 × 4–5.5 μm, wide elliptical to oval, rough to finely asperulate, hyaline.

In early summer to summer, gregarious or subcaespitose on roadsides of fields, grasslands, meadows, around farmlands, and village roadsides. Edible; cultivable Distributed in Northeast, Northwest, Inner Mongolia, and Central China.

1279 *Leucoagaricus leucothites* (Vittad.) Wasser

Pileus 5–11 cm broad, white to cream, glabrous, sometimes areolate, sometimes light grey to grey at centre. Context white. Lamellae free, white when young, beige or pinkish when mature. Stipe 5–15 cm long, 0.5–2 cm thick, cylindrical, white when young, greyish brown on middle and lower portion when mature. Annulus medial, white to creamy, persistent. Basidiospores 8–9 × 5.5–6.5 μm, elliptical, smooth, subhyaline, dextrinoid.

In summer to autumn, solitary or gregarious on ground in forests, on grassland at edge of forests or on grassland. Suspected poisonous, avoid eating it. Distributed widely in China.

1280 *Leucoagaricus nympharum* (Kalchbr.) Bon

Pileus 5–13 cm broad, plane, white to whitish, with dark grey brown to blackish brown revoluted scales. Context whitish to concolourous with pileus or lighter. Lamellae free, white to pinkish, then pale brown. Stipe 4–13 cm long, 0.4–1.5 cm thick, cylindrical, pale brown to concolourous with pileus. Annulus medial, movable, thick, beige to pale brown. Basidiospores 8–11 × 5.5–7.5 µm, elliptical to oval, not truncated at apex, smooth, hyaline, dextrinoid.

In summer, on ground or on rotten woods in thickly shaded coniferous forests. Distributed in Qinghai-Tibet Regions, China.

1281 *Leucoagaricus orientiflavus* Z.W. Ge

Pileus 3–8 cm broad, plane, waxy yellow to bright yellow, with yellowish white flocculent scales, striate at margin. Context white to slightly yellow, unchanging when injured. Lamellae free, moderately close, white. Stipe 5–10 cm long, 0.5–1 cm thick, cylindrical, yellowish white to pale yellow. Annulus superior and medial, membranous, yellow and white, movable, fugacious. Basidiospores 6.5–7.5 × 3.5–4 µm, amygdaliform in side view, narrowly oval in frontal view, smooth, hyaline, dextrinoid.

In summer, scattered on roadside and ground in forests. Distributed in Central China.

1282 *Leucocoprinus birnbaumii* (Corda) Singer

Pileus 3–8 cm broad, with yellow, sulfur yellow to yellow brown scales, radially striate at margin. Context creamy. Lamellae free, creamy yellow. Stipe 4–11 cm long, 2–5 mm thick, cylindrical, creamy yellow to yellow, obviously enlarged at base. Annulus medial to superior, creamy yellow to yellow on upper surface, pale yellow on lower surface, fugacious. Basidiospores 9–10.5 × 6–7.5 µm, oval-elliptical or amygdaliform in side view, elliptical or oval in frontal view, with an obvious germ-pore, smooth, hyaline, dextrinoid.

In summer to autumn, scattered to gregarious on ground in forests, on roadside and on soil with sawdusts in flower-pots. Reported as poisonous. Distributed widely in China.

1283 *Leucocoprinus fragilissimus* (Ravenel ex Berk. & M.A. Curtis) Pat.

Pileus 2–4 cm broad, plane, membranous, fragile, radially plicate, whitish, with yellow to light olivaceous yellow powdery squamules. Context very-thin. Lamellae free, yellowish white. Stipe 5–10 cm long, 2–4 mm thick, cylindrical, light olivaceous yellow, fragile. Annulus superior, membranous, white. Basidiospores 10–13 × 7–9 µm, elliptical to wide elliptical in side view, elliptical or oval in frontal view, smooth, hyaline, dextrinoid.

In summer to autumn, solitary to scattered on ground in forests or in grassland. Distributed in Central and South China.

1284 *Leucocortinarius bulbiger* (Alb. & Schwein.) Singer

Pileus 3–10 cm broad, flat hemispherical to plane, with an inconspicuous prominence at centre, glabrous, reddish brown, brown to pale brown, with white to dirty white felty scales, not viscid. Context white, unchanging when injured. Lamellae sinuate, white to beige. Stipe 4–9 cm long, 0.5–1 cm thick, subcylindrical, whitish, scaly, obviously enlarged at base. Basidiospores 7.5–9 × 4.5–5.5 µm, elliptical to wide elliptical, thick walled, smooth, hyaline.

In summer to autumn, on ground in forests. Edible. Distributed widely in China.

1285 *Leucopaxillus giganteus* (Sowerby) Singer

Pileus 9–40 cm broad, campanulate when young, then plane, often depressed at centre, mostly shallowly infundibuliform when mature, dirty white, greenish white or with grey yellow tinge, glabrous, zonate at maturity; margin incurved when young, then gradually straight to slightly reflexed. Context white, thick. Lamellae close, narrow, unequal, decurrent, white to milky white, beige when old. Stipe 4–8 cm long, 2–3 cm thick, stout, white to creamy, glabrous, fleshy insides. Basidiospores 6.3–8 × 4–5 µm, wide elliptical, nearly smooth, hyaline, amyloid.

In summer to autumn, on grassland, roadsides or forest edges, often forming fairy rings. Edible. Distributed in Northeast, Northwest and Inner Mongolia China.

1286 *Leucopaxillus rhodoleucus* (Romell) Kühner

Pileus 3–8 cm broad, flat to plane, white to cream, often with pinkish tint, incurved at margin. Context thick, white. Lamellae adnate to sinuate, pink to light pink. Stipe 4–8 cm long, 0.5–2 cm thick, cylindrical, pinkish white. Basidiospores 6.5–8 × 5–6 µm, wide elliptical, verruculose, amyloid.

In summer to autumn, on ground in subalpine forests. Distributed in Qinghai-Tibet Regions, China.

1287 *Leucopholiota lignicola* (P. Karst.) Harmaja

Pileus 2.5–4 cm broad, hemispherical, deep yellow brown to ochraceous, obviously scaly, with floccose scales at centre; margin incurved, with floccose to fibrillose veil when young, with fibrillose velar remnants when mature. Context thick, yellowish white to white slightly with brownish tint. Lamellae adnate, close, unequal, dirty white to pale brown. Stipe 3.5–4.8 cm long, 5–7 mm thick, cylindrical, concolourous with pileus and conspicuously floccosely scaly below annulus, enlarged and slightly darker or brown at base, pale brown or dirty brown and shallowly longitudinally ridged above annulus. Annulus superior. Basidiospores 4.5–6 × 2–2.5 µm, elliptical, thin-walled, hyaline, amyloid.

In autumn, solitary or scattered on rotten wood such as *Betula platyphylla*. Distributed in Northeast China.

1288 *Lichenomphalia hudsoniana* (H.S. Jenn.) Redhead et al.

Pileus 1–3 cm broad, flat hemispherical to plane, pale yellow to cream, glabrous, not viscid, depressed at centre, radially sulcate at margin. Context thin, whitish to pale yellow. Lamellae adnate, cream to pale yellow, distant. Stipe 3–5 cm long, 3–5 mm thick, white to dirty white, accompanied with lichenous phalli at base. Thalli 0.5–1.2 cm broad, green to deep green. Basidiospores 7–8.5 × 3.5–4.5 μm, elliptical, smooth, hyaline, amyloid.

In summer to autumn, on ground in subalpine forests, symbiotic with algae and forming thalli and basidiomata. Distributed in Qinghai-Tibet Regions, China.

1289 *Limacella ochraceolutea* P.D. Orton

Pileus 3–7 cm broad, hemispherical to flat, brownish yellow, brown to cinnamon brown at first, uneven, obtusely umbonate at centre when mature, brownish yellow, becoming lighter toward margin, glutinous, with gelatinous velar remnants at margin. Context white, unchanging when injured. Lamellae free, beige to whitish. Stipe 5–10 cm long, 0.3–1 cm thick, cylindrical, dirty white, subviscid. Annulus strongly gelatinous, often hanging at pileus margin. Basidia 25–30 × 6–7 μm. Basidiospores 4–5 × 3.5–4 μm, wide elliptical to suboval, hyaline, verruculose.

In summer to autumn, on ground in forests. Distributed in Central China.

1290 *Loreleia postii* (Fr.) Redhead et al.

Pileus 5–8 cm broad, flat semispherical at first, subinfundibuliform when mature, yellow tinged with orange red, depressed to umbilicate at centre; margin incurved when young, striate, undulate. Context whitish to pale yellow, nearly white when dried. Lamellae wide, distant, decurrent, pale yellow. Stipe 3–5 cm long, 5–8 mm thick, cylindrical, curved, tomentose, yellow brown to brown, slightly darker on middle and lower portions, with white mycelium or fibrils at base. Basidiospores 7–8 × 6.5–7.5 μm, globose, smooth, with resinous or granular contents insides, hyaline, inamyloid.

In summer to autumn, scattered on ground in forests. Distributed in Northeast China.

1291 *Lyophyllum connatum* (Schumach.) Singer

≡ *Clitocybe connata* (Schumach.) Gillet

Pileus 3–10 cm broad, convex at first, then gradually plane, sometimes slightly prominent at centre, white to nearly greyish white, rugulose-striate at margin. Context thick, white. Lamellae adnate to decurrent, unequal, close, white to pale yellow. Stipe 15 cm long, 2–5 cm thick, cylindrical, curved at lower portion, often confluent with several other stipes caespitose together, solid to soft insides. Basidiospores 5.5–7 × 2.5–3.5 μm, elliptical, smooth, hyaline.

In autumn, caespitose on ground in broad-leaved forests. Edible. Distributed in Northern, Northeast and Northwest China.

1292 *Lyophyllum decastes* (Fr.) Singer

Pileus 5–16 cm broad, flat hemispherical to plane, depressed at centre, greyish white to greyish yellow, glabrous, not viscid; margin even and incurved at first, then expanded to irregularly undulate to valvate. Context thick at centre, white. Lamellae adnate to decurrent, moderately close to close, white, unequal. Stipe 3–8 cm long, 0.7–1.8 cm thick, subcylindrical or slightly flattened, white, glabrous, solid. Basidiospores 5–7 × 4.8–6 μm, subglobose, smooth, hyaline.

In summer to autumn, caespitose on grassland or on fallen leaves in edge of broad-leaved forests or on ground rich in organic matter. Edible. Distributed in Northeast, Northern and Central China.

1293 *Lyophyllum fumosum* (Pers.) P.D. Orton

Pileus 3–6 cm broad, flat hemispherical to plane, grey to grey brown, glabrous, not viscid. Context white, unchanging when injured. Lamellae adnate to sinuate, white to dirty white, close. Stipe 4–10 cm long, 0.3–1 cm thick, cylindrical, white to greyish white, confluent with other stipes multiply with a common clumpy base. Basidiospores 5–6.5 × 4–5 μm, wide elliptical to subglobose, smooth, hyaline.

In summer to autumn, on ground in forests. Edible. Distributed widely in China.

1294 *Lyophyllum ionides* (Bull.) Kühner & Romagn.

Pileus 2–5 cm broad, flat hemispherical to plane, violet to bluish violaceous, sometimes tinted with grey to brown. Context whitish to pale brown. Lamellae sinuate, white to beige. Stipe 3–6 cm long, 3–6 mm thick, cylindrical, concolourous with pileus or lighter. Basidiospores 5–6 × 2.5–3 μm, elliptical, smooth, hyaline.

In summer to autumn, on ground in forests. Edible. Distributed widely in China.

1295 *Lyophyllum leucophaeatum* (P. Karst.) P. Karst.

Pileus 5–10 cm broad, convex, then gradually plane, slightly prominent at centre, undulate at margin, dry, felty at first, then changing to aggregately firillose, subhygrophanous, dirty white, olive grey to greyish brown, then gradually changing to pale grey brown to ochraceous brown. Context concolourous with pileus, changing to blue at first when injured, then changing to black. Lamellae sinuate to subdecurrent, close, narrow, dirty white, pale brown to greyish brown. Stipe 6–9 cm long, 1–2 cm thick, cylindrical, equal or gradually enlarged downwards, concolourous with pileus, with dark firils, dry, with white mycelium at base, with light grey pruinescence at upper portion, hollow. Basidiospores 6–8.5 × 2.5–3.5 μm, long elliptical, verrucose, hyaline.

In summer to autumn, solitary or gregarious on ground in coniferous forests, broad-leaved forests or mixed forests. Edible. Distributed in Inner Mongolia and Northwest China.

Agarics 5: Species with Names Beginning with M-O

1296 *Macrocybe lobayensis* (R. Heim) Pegler & Lodge

Pileus 3–28 cm broad, dirty white, ivory white or pale grey brown, not viscid, hemispherical or flat hemispherical at first, then gradually plane or slightly depressed at centre. Context thick, white, with an inconspicuous odour. Lamellae close, wide, sinuate, unequal, white. Stipe 7–16 cm long, 5.5–10 cm thick, enlarged and up to more than 10 cm thick at base, often multiply connected with other stipes, white, solid. Basidiospores 5–7.5 × 3.5–5 μm, oval to wide elliptical, smooth, hyaline.

Spring to summer, fasciculate to caespitose on debris in grassland or banana forests. Edible, cultivable, with a commercial name as Jinfu Mushroom. Distributed in Southern China.

1297 *Macrocystidia cucumis* (Pers.) Joss.

Pileus 2–3 cm broad, flat hemispherical to plane, red brown to dark red brown at disc, yellow at margin. Context pale yellow, with a cucumber flavour. Lamellae sinuate, white to beige. Stipe 2–4 cm long, 3–6 mm thick, cylindrical, brown to dark red brown on lower portion, becoming lighter upwards. Basidiospores 8–10 × 4–5 μm, elliptical to long elliptical, smooth, hyaline.

In summer to autumn, on ground in forests. Distributed widely in China.

1298 *Macrolepiota detersa* Z.W. Ge et al.

Pileus 8–12 cm broad, white to dirty white, with brown to pale brown fugacious patchy or flaky scales. Context white. Lamellae white to beige. Stipe 10–20 cm long, 1.5–3 cm thick, cylindrical, whitish, with concolourous squamules or subglabrous. Annulus superior, white, large, membranous, fragile. Basidiospores 14–16 × 9.5–10.5 μm, oval in side view, dextrinoid, with a germ-pore. Scales on pileus composed of a palisade of vertically arranged subcylindrical hyphae.

In summer to autumn, on ground in forests, forest edges and roadsides. Edible. Distributed in Central China.

© Central China Farmer's Publishing House 2025
Y. Li et al., *Atlas of Chinese Macrofungal Resources*, https://doi.org/10.1007/978-981-96-0502-6_6

1299 *Macrolepiota dolichaula* (Berk. & Broome) Pegler & R.W. Rayner

Pileus 6–16 cm broad, white to dirty white, with pale brown yellow to pale brownish granular squamules. Context fleshy, white. Lamellae free, white to beige. Stipe 7–24 cm long, 0.8–2.5 cm thick, cylindrical, whitish, with concolourous squamules or subglabrous. Annulus superior, membranous, white. Basidiospores 12.5–16 × 8–10.5 μm, subelliptical to oval in side view, dextrinoid, with a germ-pore. Scales on pileus composed of a palisade of subcylindrical hyphae.

In summer to autumn, on ground in forests, forest edges and roadsides. Edible. Distributed in Central China.

1300 *Macrolepiota excoriata* (Schaeff.) Wasser

Pileus 5–8 cm broad, globose at first, then plane, light earthy yellow at disc, not viscid, areolately cracked into light earthy yellow scales. Context white. Lamellae free, close, white. Stipe 5–11.5 cm long, 7–9 mm thick, cylindrical, slightly enlarged at base, white. Annulus superior, white, movable, fugacious. Basidiospores 11.2–16.3 × 7.5–9.6 μm, elliptical, smooth, hyaline, dextrinoid.

In summer to autumn, gregarious or scattered on grasslands and grasses in forests. Edible. Distributed in Northern, Northwest, Inner Mongolia, and Qinghai-Tibet Regions, China.

1302 *Macrolepiota subcitrophylla* Z.W. Ge

Pileus 7–12 cm broad, whitish, with pale brownish yellow to pale brown, pale brown or pale red brown patchy or flaky scales. Context fleshy, white. Lamellae pale yellow to cream. Stipe 10–14 cm long, 1–1.6 cm thick, cylindrical, densely with pale yellow fine squamules, subbulbous to nearly truncated at base. Annulus superior, membranous. Basidiospores 9.5–11.5 × 6.5–7.5 μm, oval-elliptical to subamygdaliform in side view, smooth, with a germ-pore, hyaline, dextrinoid. Scales on pileus composed of a palisade of vertically arranged subcylindrical elements.

In summer to autumn, solitary or scattered on ground in edges of coniferous forests. Edible. Distributed in Central China.

1301 *Macrolepiota procera* (Scop.) Singer

Pileus 7–30 cm broad, oval at first, then plano-umbonat, brown at disc, with ferruginous brown flocculent scales, dirty white at margin, not viscid. Context white. Lamellae free, close, white. Stipe 13–40 cm long, 0.8–1.5 cm thick, cylindrical, concolourous with pileus, with brown squamules, enlarged to bulbous at base. Annulus superior, fugacious. Basidiospores 14–18 × 10–12 μm, oval to wide elliptical, smooth, hyaline, dextrinoid.

In summer to autumn, solitary or scattered on grassland or ground in forest edges. Edible. Distributed widely in China.

1303 *Macrolepiota velosa* Vellinga & Zhu L. Yang

Pileus 7–9 cm broad, covered with brown to dark brown flaky or patchy scales; brown to dark brown at centre, sometimes with white to dirty white membranous velar remnants beside scales. Context fleshy, white. Lamellae white. Stipe 10–17 cm long, 0.4–1 cm thick, enlarged to subbulbous at base. Annulus superior. Volva limbate to shallowly cyathiform, membranous. Basidiospores 9–11 × 6–7.5 µm, subamygdaliform to elliptical, with a germ-pore ta apex, smooth, hyaline, dextrinoid. Cheilocystidia narrowly clavate. Scales on pileus composed of chains of suboval to subglobose cells.

In summer to autumn, on ground in tropical forests. Distributed in Southern China.

1304 *Marasmiellus candidus* (Fr.) Singer

≡ *Marasmiellus albocorticis* Singer

Pileus 0.5–3 cm broad, flat, campanulate, convex to applanate, slightly depressed at centre, membranous, white to greyish white, tomentose, striate or sulcate at margin. Context rather thin, white, odourless. Lamellae adnate to subdecurrent, distant, white, unequal, slightly forked and intervenose. Stipe 0.3–2 cm long, 1.5–4 mm thick, cylindrical, white, dull at base then dark grey brown. Basidiospores 10–17 × 3–5 µm, pip-shaped to long ellipsoid, smooth, hyaline, inamyloid.

Gregarious or caespitose on twigs or rotten wood of broadleaf trees. Distributed in South China.

1305 *Marasmiellus corticum* Singer

Pileus 0.6–4 cm broad, applanate, convex to flabelliform, depressed at centre, membranous, gelatinous when dried, white, translucent, tomentose, radially striate or sulcate. Context membranous, white. Lamellae adnate, subdistant, white, unequal. Stipe 3–9 mm long, 1–1.5 mm thick, cylindrical, eccentric, usually curved, white, tomentose, with white to yellow white basal mycelium. Basidiospores 7–10 × 4–5.5 µm, ellipsoid, smooth, hyaline.

Gregarious on decaying wood or twigs of bamboo in mixed forest. Distributed in South China.

1306 *Marasmiellus dendroegrus* Singer

Pileus 0.6–2 cm in diam., pale yellow brown to brown, applanate, plano- umbilicate or convex-umbilicate, membranous, radially sulcate. Context rather thin, pale yellow brown, odourless. Lamellae adnate, unequal, forked, tawny to orange brown or brown. Stipe 0.7–2.5 cm long, 0.4–1.8 mm thick, cylindrical, central to eccentric, yellow to tawny, hollow. Rhizomorphs yellow, well-developed. Basidiospores 5–7 × 3–4 µm, ellipsoid to pip-shaped, smooth, hyaline.

Gregarious on decayed wood in broadleaf forest. Distributed in South China.

1307 *Marasmiellus epochnous* (Berk. & Broome) Singer

Pileus 0.2–0.7 cm broad, conchate, convex, reniform, subcircular to ellipsoid, white or whitish when young, pale brown to orange grey tinged with pinkish, pruinose- tomentulose to subglabrous, sulcate. Context white to concolorous with pileus. Lamellae white, tincted with pale brown when mature, adnate or free, unequal, forked, subdistant to subcrowded. Stipe 0.3–0.5 cm long, 0.5 mm thick, eccentric to sublateral, white, pruinose-tomentose. Basidiospores 6–8 × 3.5–4.5 µm, ellipsoid, smooth, hyaline.

Gregarious on decayed sticks in broadleaf forest in summer. Distributed in South China.

1309 *Marasmiellus ramealis* (Bull.) Singer

Pileus 0.5–1.0 cm in diam., campanulate when young, applanate when mature, slightly depressed at disc, sulcate, pinkish to pale brown, darker at disc, inflexed at margin when young, gradually expanded with age. Context thin, white. Lamellae subdistant, subdecurrent, dirty white to pinkish, unequal. Stipe 0.5–1.5 cm long, 1–2 mm thick, cylindrical or curved, paler at upper part, brown to dark brown at lower part, pruinose, tomentose at base, solid. Basidiospores 8–10 × 3–4 μm, lanceolate to long ellipsoid, guttulate.

In summer, on decayed twigs or other plant remains in dark humid environments, abundant after raining. Distributed in Northeast China, Qinghai-Tibet region, etc.

1308 *Marasmiellus* cf. *quercinus* Singer

Pileus 2–2.5 cm in diam., convex to plano-convex, usually slightly depressed at disc, brown, castaneous to cinnamon, paler at central depression and margin, pale yellow, tomentose, inconspicuously striate at margin. Context thin, paler than pileus. Lamellae unequal, greyish white. Stipe 4–8 cm long, 2–3.5 mm thick, paler when young, gradually becoming pale brown to dark brown or almost concolorous with pileus, nearly blackish brown when mature, fibrous, hollow. Basidiospores 5–7 × 2.5–3.5 μm, ovoid, smooth, hyaline.

Scattered to gregarious on sticks or litters in broadleaf forest in summer. Distributed in South China.

1310 *Marasmiellus stenophyllus* (Mont.) Singer

Pileus 0.4–1.2 cm broad, white, not viscid, hemispherical to applanate, slightly depressed at centre, membranous, radially sulcate, tomentulose. Context rather thin, white. Lamellae distant, narrow, unequal, adnate to subdecurrent, white. Stipe

0.5–1.5 cm long, 1–1.5 mm thick, cylindrical, white, sometimes becoming cinnamon brown from base when mature, tomentose, with a discoid base, hollow. Basidiospores 6–8 × 3–3.5 μm, pip-shaped, smooth, hyaline, inamyloid.

In summer, caespitose or gregarious on twigs or decayed wood in broadleaf or coniferous forests. Distributed in South China.

1311 *Marasmiellus* cf. *troyanus* (Murrill) Dennis

Pileus 1.5–2 cm broad, eccentrically circular, reniform to flabelliform, white to greyish white, sometimes with tinge of greyish orange brown or flesh red, membranous, pruinose-tomentose, often sulcate. Context thin, white. Lamellae adnate, unequal, slightly intervenose, whitish. Stipe 0.2–0.3 cm long, 2–2.5 mm thick, lateral or eccentric, cylindrical, whitish to pale brown, white tomentose, solid. Basidiospores 8.5–10 × 4.5–5.5 μm, ellipsoid to pip-shaped, smooth, hyaline, inamyloid.

Solitary or subgregarious on or decayed wood or twigs in broadleaf forests. Distributed in South China, etc.

1312 *Marasmius albogriseus* (Peck) Singer

Pileus 1.5–3.3 cm broad, convex with an umbilicus to plano-umbilicate, membranous, grey brown at disc, grey white at margin, hygrophanous, radially striate or weakly sulcate, glabrous, with sparse white tomentum. Context concolorous with pileus, evanescent at margin. Lamellae adnate, subclose, obviously intervenose, white to nearly white. Stipe 3.2–5.8 cm long, 2–4.5 mm thick, cylindrical, pale brown at apex, brown at lower part, tomentose, fibrous, hollow, with white tomentose basal mycelium. Basidiospores 6–9 × 3.6–5 μm, pip-shaped, smooth, hyaline.

Gregarious to nearly caespitose on litters of bamboo groves and broadleaf forests. Distributed in South China, Central China, etc.

1313 *Marasmius albopurpureus* T.H. Li & C.Q. Wang

Pileus 1–2 cm in diam., subconvex, depressed and deeply sulcate at centre, slightly inflexed at margin, rugulose, glabrous; purple at disc when young, white to creamy white at margin, pale purple when mature, darker at sulcate folds. Context thin, white to creamy white. Lamellae adnate to subfree, white, creamy white to pale purple when young, becoming paler when mature, often underdeveloped, slightly wrinkled, distant, mostly not reaching margin of pileus. Stipe 6.3–9 cm long, 1.5–3 mm thick, cylindrical, glabrous, hollow, pale violet in upper part, gradually becoming brown downwards. Basidiospores 20–25 × 4–6 µm, subclavate to subfusoid, thin-walled, hyaline.

In summer, scattered to gregarious on debris of *Casuarina equisetifolia* or grasses. Distributed in South China.

1314 *Marasmius albostipitatus* Chun Y. Deng & T.H. Li

Pileus 1.5–2.5 cm in diam., campanulate, convex to broadly convex, orange to dark orange, broadly umbonate at centre, paler at margin, slightly striate. Context thin, white to pale yellow. Lamellae adnate, cream, subdistant. Stipe 3–6 cm long, 2–3 mm thick, cylindrical, white, glabrous. Basidiospores 7–13 × 3.5–5.6 µm, lacrymoid to ellipsoid, smooth, hyaline.

Solitary or gregarious on humus of scrub-grassland. Distributed in Central China.

1315 *Marasmius bambusinus* Fr.

Pileus 0.3–1 cm in diam., hemispherical to convex, membranous, orange brown to orange ferruginous, tomentulose, sometime sulcate when mature. Context thin. Lamellae adnate, distant to subclose, white. Stipe 4–8 cm long, 0.5 mm thick, brown, glabrous, non-insititious, with white basal mycelium. Basidiospores 17–24 × 3–4 µm, lanceolate to narrowly fusiform, smooth, hyaline, thin-walled, inamyloid.

Solitary to gregarious on fallen leaves of bamboo. Distributed in South China.

1316 *Marasmius* cf. *bellus* Berk.

Pileus 1.5–2.5 cm in diam., hemispherical to campanulate, then plano-umbilicate, membranous, pale yellow to yellow white, dry, tomentose or glabrous, entire at margin, striate. Context thin, white. Lamellae adnate, distant, narrow, pale yellow. Stipe 3–6 cm long, 1 mm thick, white at upper part, orange to brown at lower part, inconspicuously tomentose or glabrous, fibrous, hollow, non-insititious, with white basal mycelium. Basidiospores 8–12 × 3–3.5 µm, ellipsoid, apiculate, smooth, hyaline.

Gregarious or caespitose on litters of twigs in bamboo groves. Distributed in South China.

1317 *Marasmius berteroi* (Lév.) Murrill

Pileus 0.4–2 cm in diam., broadly conical, campanulate to convex, orange yellow, orange red, orange brown to ferruginous brown, dry, shout tomentulose, sulcate, slightly umbilicate at disc. Context thin, whitish to tincted with the pileus colour, tasteless or spicy. Lamellae 12–20 per cm at margin, unequal, white to pale yellow, adnate to sinuate. Stipe 2–4 cm long, 0.5–1.3 mm thick, concolorous with pileus to tinged with purple brown, paler at apex, polished, with basal mycelium. Basidiospores 10–16 × 3–4.5 µm, fusiform to lanceolate, smooth, hyaline.

In summer, gregarious on decayed sticks or litters in broadleaf forests. Distributed in South China.

1318 *Marasmius coarctatus* Wannathes et al.

Pileus 1.8–3.5 cm in diam., convex to applanate, yellow brown to greyish orange brown, glabrous or slightly striate. Context thin, pale yellow brown. Lamellae adnate, whitish to pale yellow, close; edge concolorous with lamella side face or with the colour of pileus. Stipe 3.5–6.5 cm long, 1.5–3 mm thick, cylindrical, whitish to pale yellow at apex, orange brown, yellow brown to grey brown at base, glabrous, with yellow and hackly basal mycelium. Basidiospores 5.5–9 × 3–3.5 µm, ellipsoid, smooth, thin-walled, hyaline, inamyloid.

Solitary or gregarious on decayed twigs or leaves of dicotyledons. Distributed in Northeast and Central China, etc.

1319 *Marasmius cohaerens* (Pers.) Cooke & Quél.

Pileus 2.3–2.8 cm in diam., convex to applanate, orange brown to yellow brown, darker at centre, glabrous or slightly striate, tomentulose. Context thin, white. Lamellae subfree, white to dirty white, close. Stipe 3–4 cm long, 1–2 mm thick, cylindrical, orange brown to red brown, slightly thinner at base, darker to nearly blackish brown at base, glabrous, with white hackly mycelium. Basidiospores 7–10.5 × 3.5–4.5 μm, ellipsoid, smooth, thin-walled, hyaline, inamyloid.

Solitary or gregarious on decayed twigs or leaves of dicotyledons. Distributed in South China, etc.

1320 *Marasmius coklatus* Desjardin et al.

Pileus 2–3.5 cm in diam., convex-umbonate to plano-umbonate, slightly striate, black brown or dark brown at disc, brown to pale brown at margin. Context thin, white to concolorous with pileus. Lamellae adnexed to adnate, subdistant, unequal, intervenose. Stipe 4.5–7.5 cm long, 4–5 mm thick, brown at apex, dark brown at base, with white tomentose basal mycelium. Basidiospores 8–11.5 × 4.5–7 μm, ellipsoid, smooth, hyaline, inamyloid.

Solitary or gregarious on decayed twigs or leaves of dicotyledons. Distributed in South China.

1321 *Marasmius confertus* var. *parvisporus* Antonín

Pileus 1.8–2.3 cm in diam., campanulate, oblate hemispherical, convex to applanate, orange to orange red when young, then red brown at centre, sometimes discolouring to pale orange, paler at margin, glabrous, slightly striate or not so. Context thin, white. Lamellae adnate, white, rather narrow, subclose, intervenose. Stipe 3–7 cm long, 2–3 mm thick, white near pileus, gradually changing to orange, orange brown or red brown, with white to pale yellow basal mycelium. Basidiospores 8–10 × 3–4 μm, ellipsoid, inamyloid, thin-walled.

Solitary or gregarious on decayed leaves of dicotyledons. Distributed in South China, etc.

1322 *Marasmius* cf. *corrugatus* (Pat.) Sace.

Pileus 2–4.3 cm broad, plano-hemispherical, convex to applanate or applanate with a papilla, orange, pale orange brown to grey orange, slightly darker at centre, glabrous, often slightly striate. Context thin, white to concolorous with pileus. Lamellae adnate to sinuate, subclose, white. Stipe 3–5.5 cm long, 1–3 mm thick, cylindrical, whitish at apex, gradually changing to orange brown at base, with white or pale brown basal mycelium. Basidiospores 9–11 × 3–4 μm, ellipsoid, smooth, hyaline.

On decayed wood or leaves. Distributed in South China.

1323 *Marasmius epiphyllus* (Pers.) Fr.

Pileus 0.4–1 cm in diam., convex to applanate, sometimes depressed at centre, white to creamy white, membranous, radially rugulose-striate. Context fibrous, tough. Lamellae distant, white, forked, transvenose. Stipe 1.5–3 cm long, 1–2 mm thick, fibrous, whitish at upper part, pale red brown downwards. Basidiospores 9.5–11 × 3–4.5 μm, long ellipsoid, smooth, hyaline, inamyloid.

Solitary or scattered on decayed twigs or leaves in autumn. Distribute in North China, Northeast China, South China, etc.

1324 *Marasmius floriceps* Berk. & M.A. Curtis

Pileus 1–3 cm in diam., plano-hemispherical, convex to applanate, sulcate, rugulose at centre, orange, orange red to orange brown, darker at centre. Context thin. Lamellae adnexed, subclose, unequal, narrow, white. Stipe 2–4.5 cm long, 1–1.5 mm thick, cylindrical, hollow, nearly white, yellow white to slightly greenish yellow at apex, gradually changing to orange brown towards base, with white to pale yellow basal mycelium. Basidiospores 6.8–9.2 × 3–3.4 μm, ellipsoid, thin-walled, hyaline, inamyloid.

Solitary or gregarious on decayed leaves or branches of dicotyledons. Distributed in Northeast China, North China, South China, etc.

1325 *Marasmius galbinus* T.H. Li & Chun Y. Deng

Pileus 0.8–1.5 cm in diam., conical when young, campanulate to convex when mature, rugulose at centre, distinctively striate, hygrophanous, pale yellowish white to slightly greenish white, whitish at disc and striae. Context thin, yellow white. Lamellae 1–2 mm wide, subfree, subdistant, unequal. Stipe 1.5–3 cm long, 0.5–1 mm thick, cylindrical, hollow, glabrous, non-insititious, nearly white at upper part, pale yellow to pale yellow brown at lower part, with white to pale yellow basal mycelium. Basidiospores 14–16 × 4–5 µm, subfusiform, often curved, smooth, hyaline, inamyloid.

Gregarious on ground under dicotyledons or weeds. Distributed in South China.

1326 *Marasmius graminum* (Lib.) Berk.

Pileus 0.4–0.6 cm in diam., hemispherical or campanulate, umbilicate depressed and papillate at umbilicus, dirty white to pale yellow, then yellow brown, deep orange to brown, inconspicuously tomentose or not so, radially deeply striate or sulcate. Context thin, white, tasteless. Lamellae 7–9 per cm at margin, unequal, yellow, free, with a collarium. Stipe 0.2–1 cm long, 0.5–1 mm thick, slender, pale yellow at upper part when young, orange brown to dark brown entirely or at lower part when mature. Basidiospores 8–12 × 3.5–4.5 µm, slenderly pip-shaped, smooth, hyaline.

Scattered on herbaceous plants and debris in broadleaf forests. Distributed in Northeast China, Inner Mongolia, South China, etc.

1327 *Marasmius haematocephalus* (Mont.) Fr.

Pileus 0.5–2.5 cm in diam., campanulate at first, then convex to plano-umbonate, red brown to purplish red brown, dry, densely tomentulose, slightly striate or sulcate. Context thin. Lamellae sinuate to free, subdistant, white when young, then changing to yellowish white, with very few lamellulae. Stipe 3–5.5 cm long, 0.5–1 mm, deep brown or dark brown, yellow white at apex, cartilaginous, slightly enlarged and discoid at base, with white basal mycelium. Basidiospores 16–26 × 4–5.5 μm, long fusiform, smooth, hyaline.

Gregarious on decayed twigs or leaves in broadleaf forests. Distributed in Northwest China, Central China, South China, etc.

1328 *Marasmius* cf. *helvolus* Berk.

Pileus 1–2 cm broad, convex when young then applanate, umbonate, membranous to subleathery, pale grey brown to deep brown, radially striate or sulcate, often inflexed at margin. Context thin, white to yellow white. Lamellae adnate, subclose, white to yellowish white, unequal. Stipe 4–6 cm long, 1–2 mm thick, yellow white at apex, gradually changing to brown downwards, fibrous, with white tomentum and mycelium at base. Basidiospores 12–15 × 3–4.5 μm, subfusoid, smooth, hyaline.

Scattered to gregarious on litters in broadleaf forests. Distributed in South China.

The Chinese samples slightly differ from the South American materials of *Marasmius helvolus* by its darker pileus according to the relevant description. The Chinese samples are treated here as a reference species.

1329 *Marasmius hinnuleus* Berk. & M.A. Curtis

Pileus 0.8–1 cm in diam., convex to applanate, brown to dull brown at centre, greyish orange-brown to greyish yellow-brown at margin, striate, glabrous. Context thin. Lamellae adnexed to free, white, subdistant, with few lamellulae. Stipe 2.5–3.5 cm long, 0.5–1 mm thick, cylindrical, white at apex, gradually changing to pale brown, ochre to nearly black brown, with yellowish basal mycelium. Basidiospores 8–11 × 3.8–5 μm, nearly long ellipsoid, smooth, hyaline.

Solitary or gregarious on decayed leaves in broadleaf forests. Distributed in South China, etc.

1330 *Marasmius hygrocybiformis* Chun Y. Deng & T.H. Li

Pileus 1.5–3 cm in diam., applanate, orange red to red orange, slightly striate, bright and translucent, glossy, similar to *Hygrocybe* sp. Context thin, white to pale brown. Lamellae adnate, yellow white, wax yellow to pale yellow, translucent, subdistant, with the colour of pileus at edge. Stipe 3–5 cm long, 2–3 mm thick, slender, cylindrical, pale green, translucent, glabrous, with orange yellow basal mycelium. Basidiospores 8.5–10.5 × 4–5.7 μm, ellipsoid, smooth, hyaline.

Solitary or gregarious on decayed twigs or leaves of dicotyledon. Distributed in Qinghai-Tibet region.

1331 *Marasmius* cf. *hymeniicephalus* (Speg.) Singer

Pileus 0.7–4 cm in diam., oblate hemispherical, convex to nearly plane, sometimes slightly umbilicate or slightly papillate at centre, membranous, creamy white, sometimes slightly tincted with eggshell colour, striate or sulcate. Context very thin, white. Lamellae adnate, white to yellowish, subclose, forked and intervenose. Stipe 1.2–5 cm long, 1–2.5 mm thick, cylindrical, white to yellow white at apex, gradually changing to creamy yellow to yellow brown downwards, white and short tomentose, fibrous, hollow, with dense white basal tomentum. Basidiospores 6–7 × 3–3.5 μm, ellipsoid, smooth, hyaline.

Caespitose to gregarious on rotten or decayed twigs or leaves in broadleaf forests. Distributed in South China, North China, etc.

1332 *Marasmius maximus* Hongo

Pileus 3.5–10 cm in diam., campanulate or hemispherical at first, then applanate, often slightly umbonate at centre, subhygrophanous, radially sulcate to plicate, tawny to brown, often deep brown at disc, more or less discolouring to pale brown or pale yellow outwards, sometimes to nearly yellowish white, even nearly white when dried. Context thin, subfleshy to subleathery. Lamellae 2–7 mm wide, adnate, emarginate to free, subdistant, unequal, paler than pileus. Stipe 5–9 cm long, 2–3.5 mm thick, equal, tough, pruinose at upper part, solid. Basidiospores 7–9 × 3–4 µm, subfusoid to ellipsoid, smooth, hyaline, inamyloid. Cheilocystidia 16–30 × 6.5–10 µm, clavate or irregular.

Scattered to gregarious on decayed leaves in forest or grassland. Distributed widely.

1333 *Marasmius* cf. *neosessilis* Singer

Pileus 0.4–1.8 cm broad, flabelliform or pleurotoid, orange brown or pale red brown, discolouring to flesh-colour when old, membranous, not viscid, finely short tomentulose or glabrous, margin entire, sparsely and shallowly sulcate. Context very thin, white, odourless or garlic. Lamellae distant, ridged or veined, erected, yellow white or creamy yellow, unequal or forked, narrow, intervenose. Stipe about 0.2 cm long, lateral or reduced, paler or nearly concolorous with pileus, tomentose, solid. Basidiospores 6–9 × 3.5–5 µm, ellipsoid or broad ellipsoid, smooth, hyaline.

On bark, decayed wood or branches of broadleaf trees (such as *Cinnamomum camphora* (L.) Presl). Distributed in South China.

1334 *Marasmius nigrodiscus* (Peck) Halling

Pileus 3–10 cm in diam., obtusely convex at first, then gradually expanded, pale yellow, grey yellow, orange yellow to yellow brown, darker at centre, yellow brown to brown, often dry, hygrophanous when moist. Context thin, white. Lamellae adnate to subfree, subdistant, unequal, white then pale yellow. Stipe 6–10 cm long, 0.6–1 mm thick, whitish or concolorous with pileus, hollow, with white basal mycelium. Basidiospores 6–10 × 3–4.5 µm, ellipsoid to lacrymoid, smooth, thin-walled, hyaline.

On ground of coniferous or mixed broadleaf-conifer forests. Distributed in Central China, South China, etc.

1335 *Marasmius* cf. *occultatiformis* Antonin et al.

Pileus 1–2.5 cm in diam., hemispherical, convex to applanate, red brown at centre, orange brown at margin. Context thin, whitish. Lamellae adnate, white, subclose. Stipe 4–7 cm long, 2–3 mm thick, cylindrical, white at apex, translucent, gradually changing to red brown, dark brown, non-insititious, with white basal mycelium. Basidiospores 6.5–8 × 3–4 µm, ellipsoid, smooth, hyaline.

Solitary or gregarious on humus of *Picea asperata* Mast., *Abies fabri* (Mast.) Craib or shrub forests. Distributed in Qinghai-Tibet region.

1336 *Marasmius* cf. *ochroleucus* Desjardin & E. Horak

Pileus 1.1–1.5 cm in diam., convex to plano-convex, yellow to cream, paler at margin, papillate at centre, striate, hygrophanous. Context thin, white. Lamellae adnate, white, rather narrow. Stipe 3–4.5 cm long, 1–2 mm thick, white at apex, transparent, gradually changing to yellow brown, with white to yellow basal mycelium. Basidiospores 9.8–11.5 × 3.6–4.2 µm, long ellipsoid, curved, smooth.

Solitary or gregarious on leaves and decayed twigs of monocotyledons or dicotyledons. Distributed in Northwest China, etc.

1337 *Marasmius oreades* (Bolton) Fr.

Pileus 2.5–5 cm in diam., plano-hemispherical when young, gradually expanded with age, pale flesh-colour to tawny, slightly umbonate at centre, margin glabrous, transparently striate when moist. Context thin, whitish to concolorous with pileus. Lamellae white to dull white, adnate, distant, unequal. Stipe 3–7 cm long, 3–5 mm thick, cylindrical, yellowish white to brown, tomentose-furfuraceous, solid. Basidiospores 7.5–9.5 × 3–3.5 µm, ellipsoid, smooth, hyaline.

On grassland, roadside, field, forest, etc. Often forming fairy rings. Edible and medicinal. Distributed in Northeast China, North China, Inner Mongolia, Central China, etc.

1338 *Marasmius purpureostriatus* Hongo

Pileus 1–2.5 cm in diam., campanulate to hemispherical, umbilicate, papillate at centre, with purple brown or pale brownish purple radial sulci, paler when old. Context thin, dirty white. Lamellae subfree, dirty white to creamy white, distant, unequal. Stipe 4–11 cm long, 2–3 mm thick, cylindrical, dull white at upper part, gradually brown towards base, finely tomentulose, often coarsely white strigose at base, hollow. Basidiospores 22.5–30 × 5–7 µm, long clavate, smooth, hyaline.

On decayed twigs or leaves in broadleaf forests during summer to autumn. Distributed in Northeast China, North China, South China, etc.

1339 *Marasmius rotalis* Berk. & Broome

Pileus 1.5–7.5 cm in diam., hemispherical at first, then convex, with a small papilla at centre, white, yellowish white to pale brown, darker at centre, striate or sulcate. Context thin, concolorous with pileus. Lamellae adnate, with a collarium. Stipe 2–2.5 cm long, 0.5–1 mm thick, cylindrical, hollow, dark brown. Rhizomorphs present, black. Basidiospores 7–9 × 3–4 µm, ellipsoid, smooth, hyaline.

On decayed leaves. Distributed in South China.

1340 *Marasmius* cf. *setulosifolius* Singer

Pileus 0.8–3 cm in diam., campanulate to hemispherical or convex when young, then plano-umbonate, rugulose at centre, pale yellow, orange cinnamon to pale ferruginous, inviscid, glabrous. Context thin, white. Lamellae adnate, with 10–12 complete lamellae and some lamellulae, narrow, paler than pileus. Stipe 2.5–4 cm long, 0.7–2 mm thick, cylindrical, whitish, pale cinnamon, ochre-brown to pale brown; base radially tomentose or strigose, white, cinnamon to pale yellow brown. Basidiospores 7.5–10 × 3–4 µm, ellipsoid to subfusiform, apiculate, smooth, hyaline. Cheilocystidia same to pleurocystidia, 15–30 × 3.8–8 µm, clavate to cylindrical, as broom-like cells of *Siccus*-type. Pileipellis cells, 25–38 × 13–20 µm, broom-like.

Solitary on decayed twigs or leaves in mixed or broadleaf forest. Distributed in South China.

The Chinese samples are very similar to *M. setulosifolius* Singer from South America. However, *M. setulosifolius* differs in its distinct ferruginous brown pileus, slightly larger basidiospores, smaller broom cells in pileipellis (15–25 × 9–10 µm). The Chinese samples are treated here as a reference species of *M. setulosifolius*.

1341 *Marasmius siccus* (Schwein.) Fr.

Pileus 1–2 cm in diam., hemispherical, convex to applanate, orange yellow, ochre yellow, orange to dark orange, depressed at centre, papillate at central depression, striate. Context thin, white. Lamellae 1–1.5 mm wide, sinuate to subfree, white, subdistant, with lamellulae or not; edge tincted with the colour of pileus, but inconspicuously in some specimens. Stipe 2–5 cm long, 0.5–1.5 mm thick, cylindrical, white at upper part, gradually changing to dark castaneous to black downwards, glabrous, with lacquered sheen, with white to yellow white basal mycelium. Basidiospores 16–21 × 3–4 μm, lanceolate, often curved, smooth, white.

Solitary or gregarious on decayed leaves in broadleaf forest in summer. Distributed in Northeast China, North China, Northwest of China, Central China, etc.

1342 *Marasmius subabundans* Chun Y. Deng & T.H. Li

Pileus 0.8–1 cm in diam., convex to applanate, glabrous, dry, yellow white, orange white, pale orange to orange yellow, cream to creamy white at margin. Context thin, yellow white. Lamellae about 0.1 cm wide, adnate, subclose, unequal, white to yellow white. Stipe 3–5 cm long, 0.5–1 mm thick, cylindrical, translucent and white at apex, pale brown to brown at base, with white to yellowish white basal mycelium. Basidiospores 7–9 × 3–4.5 μm, ellipsoid to subfusiform, smooth, hyaline.

Gregarious on decayed leaves of dicotyledon. Distributed in South China.

1343 *Marasmius suthepensis* Wannathes et al.

Pileus 0.8–2.2 cm broad, convex, then gradually expanded, orange brown to pale orange at centre, fading to pale yellow, orange white to pale yellow at margin, papillate or not, smooth to weakly sulcate, glabrous. Context thin, white. Lamellae adnate to free, subclose, edge tincted with the colour of pileus. Stipe 2–5.5 cm long, 1 mm thick, cylindrical, yellow white at apex, red brown at base, with yellow basal mycelium. Basidiospores 10–13 × 3.1–4 μm, ellipsoid, smooth, hyaline.

Solitary or gregarious on decayed leaves and wood of dicotyledon. Distributed in Central China, South China, etc.

1344 *Marasmius umbo-albus* Chun Y. Deng & T.H. Li

Pileus 2–4 cm broad, convex, broadly convex to applanate, slightly and broadly umbonate, white at centre, yellowish white elsewhere, hygrophanous, glabrous, inconspicuously to rather obviously striate. Lamellae adnate, intervenose, white to creamy white. Stipe 6–10 cm long, 2–5 mm thick, cylindrical, hollow, glabrous, pale yellow, orange white, orange brown. Basidiospores 5.5–9 × 3–4 μm, fusiform to subcylindrical, smooth, hyaline.

Solitary or gregarious on litters in mixed forests. Distributed in Central China.

1345 *Megacollybia clitocyboidea* R.H. Petersen et al.

Pileus 5–13 cm broad, campanulate when young, plane to upturned at maturity, slightly depressed at centre, mouse grey to dark black, often fading to yellow brown at margin, often rough at disc, often split at margin at maturity. Context very thin, dirty white. Lamellae adnate to subdecurrent, whitish, greyish white or pinkish at age, distant or very distant. Stipe 5–10 cm long, 1–2 cm thick, cylindrical, dirty white on upper portion, greyish on lower portion, enlarged at base, solid, tomentose or pruinose, with white to greyish white rhizomorphs at base when young. Basidiospores 6–9 × 5.5–7 μm, broadly elliptical, smooth, hyaline.

In summer to autumn, on rotten wood or ground in forests. Distributed in Northeast, and Qinghai-Tibet Regions, China.

1346 *Melanoleuca brevipes* (Bull.) Pat.

Pileus 2.5–8.5 cm broad, convex, then gradually plane, glabrous, dry, dark grey to almost black at first, then gradually changing to grey to dark brown. Context thick, white. Lamellae sinuate, close, white. Stipe 2–4 cm long, 0.8–1.2 cm thick, short clavate, yellow brown to greyish brown, lighter at apex. Basidiospores 6.5–9.5 × 5–6.5 μm, elliptical to broadly elliptical, verrucose, hyaline, amyloid.

In summer to autumn, solitary or gregarious on grasslands. Distributed in Northeast and Northwest China.

1347 *Melanoleuca leucopoda* X.D. Yu

Pileus 2–4 cm broad, slightly prominent at disc at first, then plane, cinnamon, darker at disc, viscid when wet, tomentose. Context white to cream. Lamellae subsinuate, close, white, with lamellulae. Stipe 4–8 cm long, 4–7 mm thick, cylindrical, white, cinnamon at base, tomentose, slightly enlarged at base. Basidiospores 10–14 × 6–8 μm, long ellipsoid, with elongated warts, hyaline, amyloid. Cystidia fusiform, thick walled, encrusted with crystals at apex.

In summer to autumn, scattered on ground in open forests. Distributed in Northeast China.

This species is similar to *M. cognata*, but the latter has larger pileus, and different shape of basidiospore ornamentation.

1348 *Melanoleuca porphyropoda* X.D. Yu

Pileus 3–4 cm broad, plane, undulate at margin, dark yellow brown, slightly darker at disc, with white tomentum. Context white to cream. Lamellae decurrent, closer, white, with lamellulae. Stipe 5–6 cm long, 5–7 mm thick, cylindrical, violaceous, with white tomentum, slightly enlarged at base. Basidiospores 8–12 × 4.5–8 μm, elliptical, verrucose, hyaline, amyloid.

In summer to autumn, scattered on grassland. Edibility unknown. Distributed in Northeast China.

This species is similar to *M. graminicola* and *M. melaleuca*, but the stipe of the latter two species are not violaceous.

1349 *Melanoleuca verrucipes* (Fr.) Singer

Pileus 4–12 cm broad, convex, then gradually plane, slightly prominent at centre, whitish, yellow brown at disc, glabrous, incurved at margin at first, then plane. Context tastes mild, white. Lamellae close, dirty white to dirty yellow. Stipe 3–8 cm long, 0.7–1.5 cm thick, cylindrical, concolourous with pileus, covered with brown to dark brown granular warts. Basidiospores 8–10 × 4.5–5.5 μm, elliptical, verrucose, hyaline, amyloid.

In summer to autumn, gregarious or scattered on ground in mixed forests or on grasslands at forest edges. Edible. Distributed in Northeast and Qinghai-Tibet Regions, China.

1350 *Melanophyllum haematospermum* (Bull.) Kreisel

Pileus 0.5–3.5 cm broad, convex to subplane, slightly droopy to plane at margin, pale yellow white to pale white brown at disc, slightly darker at margin, pale brown to dirty white brown, even, estriate, not hygrophanous, densely with brown, dark brown to nearly black finely granular to powdery scales. Context thin, whitish to concolourous with pileus. Lamellae close, unequal, adnate to slightly decurrent, pale yellowish brown, with pale brownish yellow, salmon brown or purplish red to vinaceous red tinge. Stipe 2–6 cm long, 1.5–4 mm thick, cylindrical, lavender red, purplish red to violaceous red brown, covered with brown to violaceous red brown fagacious pulverulent scales in lower portion. Annulus superior, fagacious. Basidiospores 4–5 × 2–3 μm, long ellipsoid, verrucose.

On humus. Distributed in Northeast and Central China.

1351 *Micromphale foetidum* (Sowerby) Singer

Pileus 3–4.6 cm broad, plane, often depressed at centre, dark brown to pale brown at disc, slightly lighter at margin, sulcate, glabrous, hygrophanous. Context thin, pale grey brown to concolourous with pileus. Lamellae adnate to adnexed, very close, pale yellow brown. Stipe 3–4 cm long, 2–3 mm thick, pale yellow brown at apex, brown or dark brown at base, hollow, with yellow white tomentum and mycelium at base. Basidiospores 7.5–9 × 3–4 µm, pip-shaped, smooth, hyaline.

Scattered on debris in broad-leaved forests. Distributed in Southern China.

1352 *Micropsalliota furfuracea* R.L. Zhao et al.

Pileus 2–3.5 cm broad, obtusely conical at first, then plane with a prominent centre, dirty white to slightly brownish, striate at margin, with dense brown appressed scales at centre, with furfuraceous squamules at margin. Context white, changing to red brown to dark brown when injured or at age. Lamellae free, unequal, close, brownish brown to tan. Stipe 2.5–3.5 cm long, 2.5–3.5 mm thick, subequal, hollow, fibrous, white to pale yellow at first, red brown when injured, dark brown to dark purplish brown in later stage. Annulus superior, single. Basidiospores 6–7.5 × 3–4 µm, elliptical, smooth, brown.

Gregarious or caespitose on ground in broad-leaved forests. Distributed in Southern China.

1353 *Mycena acicula* (Schaeff.) P. Kumm.

Pileus 0.2–0.9 cm broad, hemispherical at first, then gradually changing to broadly conical, pale orange red to orange yellow, becoming lighter toward margin; sulcate at margin. Context thin, creamy yellow to pale orange yellow, with an inconspicuous odour. Lamellae adnate to sinuate, entire at edge. Stipe 1–5 cm long, 1–2 mm thick, cylindrical, subequal, hollow, slightly viscid, lemon yellow at first, gradually white at base, with white pruinescence on upper portion, white fibrils at base. Basidiospores 8.5–12 × 3–4 μm, long ellipsoid to subfusiform, smooth, hyaline, amyloid.

In summer to autumn, solitary or scattered on litter. Distributed in Northeast China.

1354 *Mycena adonis* (Bull.) Gray

Pileus 5–15 cm broad, subconical, papillate at centre, gradually plane at age, subglabrous, shallowly sulcate or translucently striate at margin, pinkish salmon flesh-coloured, pink or bright red, faded at age, sometimes whitish. Context thin, whitish to concolourous with pileus. Lamellae adnate to decurrent, pale pink to white. Stipe 3.5–5 cm long, 1.5–2 mm thick, cylindrical, hollow, brittle, pale pink or pink on upper portion, gradually white downward, subtranslucent, hygrophanous, finely fibrillose. Basidiospores 7–9 × 3.5–5 μm, elliptical, multi-guttulate, thin-walled, with a conspicuous germ-pore, hyaline, amyloid.

On rotten wood in mixed forests. Distributed in Northeast China.

The macroscopic characteristics of this species is similar to *M. acicula*, but the pileus of this species is slightly pink, and the stipe is usually pale pink or whitish; while the pileus of the latter species is orange red, and the stipe is lemon yellow or whitish.

1355 *Mycena alphitophora* (Berk.) Sacc.

Pileus 0.3–0.6 cm broad, convex at first, then gradually changing to campanulate, with white pruinescence, striate, light grey at first, gradually fading to dirty white in later stage. Context thin, inconspicuous smell and taste. Lamellae free to slightly decurrent, distant, narrow, white. Stipe 2–3.5 cm long, 1–2 mm thick, cylindrical, gradually enlarged at base, densely covered with white tomentum, then gradually with white pruinescence. Basidiospores 7.5–9.5 × 4–5 μm, elliptical, smooth, hyaline, amyloid.

In summer to autumn, solitary to scattered on debris. Distributed in Northeast and Central China.

1356 *Mycena auricoma* Har. Takah.

Pileus 1–3 cm broad, hemispherical to plane, yellow to brownish yellow, lighter at margin. Context thin. Lamellae beige to pale yellow. Stipe 2–3 cm long, 1–3 mm thick, cylindrical, white to beige, hollow. Basidiospores 5–7 × 3–4 μm, elliptical to broadly elliptical, smooth, hyaline, amyloid. Cheilocystidia subfusiform, slender at apex.

In summer to autumn, on rotten wood in forests. Distributed in Central China.

1357 *Mycena capillaripes* Peck

Pileus 1–2.5 cm broad, campanulate at first, gradually plane when mature, convex, prominent at apex, glabrous, sulcate, grey brown. Context very-thin. Lamellae adnate, unequal, distant, dirty white or grey, pinkish at edge. Stipe 4–6 cm long, 1–2 mm thick, cylindrical, concolourous with pileus or slightly lighter, sometimes brownish at base, cartilaginous, hollow, subequal or gradually thicker toward base, glabrous, occasionally with powdery squamulae at apex. Basidiospores 8–11 × 4–6.5 μm, elliptical, smooth, hyaline, amyloid.

In autumn, scattered or gregarious on ground in coniferous forests. Distributed in Northeast China.

1358 *Mycena corynephora* Maas Geest.

Pileus 5–8 cm broad, campanulate to umbelliform or hemispherical, subglabrous or pruinose, pure white, estriate. Context very-thin, white, hygrophanous. Lamellae distant, narrow, adnate. Stipe slender, 7–15 mm long, 0.5–1.5 mm thick, cylindrical, curved. Basidiospores 6–8 × 8–9.5 μm, globose or subglobose, smooth, hyaline, amyloid. Pileipellis composed of a layer of globose cells in 6–27 × 8–31 μm; globose cells with verrucous spines of varying lengths, hyaline, often with a short pedicel-like base. Clamp connection absent.

On bark of broad-leaved trees (*Salix*). Distributed in southern China.

1359 *Mycena epipterygia* (Scop.) Gray

Pileus 1–2.5 cm broad, conical to hemispherical at first, plane in later stage, sometimes slightly prominent at centre, glabrous, grey brown to khaki, striate at margin when wet, viscid. Context thin, whitish to concolourous with pileus. Lamellae adnate to sinuate, subdistant, whitish. Stipe 5–8.5 cm long, 1–2 mm thick, yellow green, covered with fine fibrillose hairs on lower portion. Basidiospores 8.5–10.5 × 5–6 μm, oval to elliptical, smooth, hyaline, amyloid.

In summer, caespitose or gregarious on rotten wood of broad-leaved trees in mixed forests. Distributed in Northeast China.

1360 *Mycena galericulata* (Scop.) Gray

Pileus 2–5 cm broad, campanulate when young, gradually plane when mature, translucent, sulcate or obviously rugulose, darker when young, then lead grey, dark at disc, whitish and occasionally slightly lacerated at margin. Context translucent, thin, not an obvious odour. Lamellae subclose, white, unequal, adnate to sinuate, slightly decurrent when young, sometimes furcate or interveined. Stipe 4–8 cm long, 2–5 mm thick, cylindrical or flat, dark grey when young, grey to greyish white when mature, glabrous, hollow, cartilaginous, with white hairy mycelium at base. Basidiospores 9.5–12 × 7.5–9 μm, broadly elliptical, smooth, hyaline, amyloid.

In early summer to autumn, on stumps, rotten wood or dead branches of broad-leaved or coniferous trees in forests. Distributed in Northeast, Inner Mongolia and Central China.

1361 *Mycena haematopus* (Pers.) P. Kumm.

Pileus 2.5–5 cm broad, conical when young, gradually changing to campanulate, striate; dark red when young, slightly lighter when mature, dark at disc, lighter and often lacerated to regularly serrate at margin; with fine pruinose granules when young, then glabrescent, exuding sanguineous latex when injured. Context thin, white to vinaceous. Lamellae adnate to subsinuate, white to greyish white, sometimes with dark red spots, close. Stipe 3–6 cm long, 2–3 mm thick, cylindrical or flattened, equal, concolourous with pileus or slightly lighter, covered with white fine farinaceous granules, hollow, fragile, with white hairy mycelium at base. Basidiospores 7.5–11 × 5–7 μm, broadly elliptical, smooth, hyaline, amyloid.

In early summer to autumn, often fasciculate on very decayed wood of broad-leaved trees. Distributed in Northeast and Central China.

1362 *Mycena luteopallens* Peck

Pileus 0.8–1.5 cm broad, hemispherical or convex, often with a minute papilla at centre when mature, glabrous, translucently striate or slightly undulate at margin; bright orange yellow to bright yellow when young, then fading to yellowish white. Lamellae adnate, distant, wide, pale yellow to slightly pinkish, whitish at edge. Stipe 5–9 cm long, 1–2 mm thick, cylindrical, equal, hollow, glabrous to tomentulose, concolourous with pileus to orange yellow on upper portion, becoming pale yellow gradually to base, densely with basal mycelium. Basidiospores 8–9 × 4–4.5 μm, elliptical or oval, smooth, hyaline, amyloid.

Often on nut shells of *Juglans*. Distributed in Northeast China.

1363 *Mycena metata* (Fr.) P. Kumm.

Pileus 1–2 cm broad, conical to campanulate, sometimes upturned at margin, umbonate at centre, glabrous or finely fibrillose, striate at margin, hygrophanous, beige brown to nearly red brown, darker at centre, red brown or wine red, gradually lighter to margin, finely fibrillose at margin. Context thin, pale brown, hygrophanous. Lamellae adnate to subdecurrent, yellow brown to light red brown, glabrous at edge. Stipe 5–8 cm long, 1–2 mm thick, cylindrical, glabrous, pale brown to brown, darker at base, hollow, brittle. Basidiospores 7–9 × 3–4 µm, long ellipsoid to pip-shaped, smooth, hyaline, amyloid. Pileipellisfine with finger-like warts on hypha surface.

Gregarious or scattered on decayed leaves or pine needles in deciduous or coniferous forests. Distributed in Northeast China.

1364 *Mycena pelianthina* (Fr.) Quél.

Pileus 1.5–5 cm broad, hemispherical to plane, glabrous, viscid when wet, hygrophanous, purplish brown, radially sulcate-striate at margin, translucent, lavender, sometimes slightly with fleshy pink tinge. Context thin, fleshy pink, hygrophanous, often with a conspicuous odour of radish, with stronger odour when mature. Lamellae decurrent, unequal, distant, violaceous or deep purplish brown, undulate or serrate at edge. Stipe 2–5 cm long, 2–4 mm thick, hollow, tubular, central, concolourous with pileus or slightly darker on upper portion, fibrously striate, with white mycelium at base. Basidiospores 6–7.5 × 3–4 µm, elliptical, smooth.

Gregarious or solitary on humus under broad-leaved forests. Distributed in Northeast China.

This species is similar to *M. pura*, but the shorter stipe, apiculate pleurocystidia, and obvious refractive content are different.

1365 *Mycena picta* Harmaja

Pileus 2.5–5.5 cm broad, cylindrical at first, then campanulate, not expanded, depressed at centre, radially sulcate, glabrous, dry, radially translucently striate when wet, dark olivaceous grey to olivaceous brown, pale yellow and slightly incurved at margin. Context thin, pale brown. Lamellae adnate to slightly decurrent, dark green brown, red yellow at edge. Stipe 3–5 cm long, about 1 mm thick, linear, glabrous, pale brown to pale yellow brown. Basidiospores 5.5–10 × 3.2–5 µm, elliptical to long ellipsoid, smooth, hyaline, amyloid.

In autumn, scattered on ground in mixed forests. Distributed in Northeast and Qinghai-Tibet Regions, China.

1366 *Mycena polygramma* (Bull.) Gray

Pileus 2–4 cm broad, conical at first, then campanulate to plane, prominent at centre, glabrous, grey to grey brown, radially striate. Context thin, pale grey. Lamellae free, distant, whitish. Stipe 6–10 cm long, 2–3 mm thick, cylindrical, equal, with rhizomorphs at base, glabrous, lighter than pileus, conspicuously longitudinally striate. Basidiospores 9.5–12 × 6.5–8.5 µm, broadly elliptical, smooth, hyaline, amyloid.

In summer to autumn, on debris in broad-leaved forests. Distributed in Northeast and Northern China.

1367 *Mycena pura* (Pers.) P. Kumm.

Pileus 2.5–5 cm broad, hemispherical when young, then plane to slightly reflexed at margin, striate; purplish red when young, slightly lighter when mature, darker at disc; margin lighter, lacerated to regularly serrate. Context thin, greyish violaceous. Lamellae close, adnate to subsinuate, often interveined, unequal, white to greyish white, sometimes lavender. Stipe 3–6 cm long, 3–5 mm thick, cylindrical or flattened, equal or slightly thicker downwards, concolourous with pileus or slightly lighter, glabrous, hollow, cartilaginous, with white hairy mycelium at base. Basidiospores 6.5–8 × 4–5 μm, elliptical, smooth, hyaline, amyloid.

In summer to autumn, scattered on ground in mixed or coniferous forests. Distributed in Northeast, Inner Mongolia, Northwest, and Qinghai-Tibet Regions, China.

1368 *Mycena rosea* Gramberg

Pileus 1–2.5 cm broad, hemispherical when young, convex when mature, prominent at centre, glabrous or radially fibrillose, translucently striate, grey brown to yellow brown, sometimes slightly with violaceous tinge, serrated at margin. Context hygrophanous, grey brown, thin. Lamellae white when young, grey brown when mature, sometimes slightly tinged with pinkish brown, broadly adnate to subdecurrent, even at edge, violaceous to red brown. Stipe 2–4.5 cm long, 1.5–2.5 mm thick, cylindrical, often curved, glabrous, light grey brown with a little purplish brown tinge, light at apex, with fine white pruinescence, hollow, darker on lower portion, with white bristles at base. Basidiospores 9–13 × 6–8 μm, elliptical, smooth, hyaline, amyloid.

Gregarious on decayed wood or debris of broad-leaved trees. Distributed in Central China.

The morphology of this species is similar to that of *M. pura* in the similar pinkish violaceous fruit bodies, but the pileus of the latter species is subplane or convex, lacking a prominent pileus centre, while the pileus of this species has an obviously prominent centre.

1369 *Mycena sanguinolenta* (Alb. & Schwein.) P. Kumm.

Pileus 0.5–1.3 cm broad, conical to campanulate, purple red brown, lighter at margin, radially striate when wet. Context thin, whitish to concolourous with pileus. Lamellae adnate, white or light red, slightly tinged with reddish brown. Stipe 2.5–5 cm long, 0.5–1 mm thick, concolourous with pileus, with white hairs at base, exuding red latex when injured. Basidiospores 7.5–9.5 × 4–4.5 μm, elliptical, smooth, hyaline, amyloid.

In spring to autumn, on debris in broad-leaved and mixed forests. Distributed in Northeast, Northern, Northwest, Central China, and Southern China.

1370 *Mycena stylobates* (Pers.) P. Kumm.

Pileus 7–12 cm broad, viscid, conical to convex, plane when mature, translucently deeply sulcate-striate, greyish white, pruinose, brownish grey at disc. Context thin, white to translucent. Lamellae distant, decurrent, easily split from stipe, often forming a pseudoannulus on stipe. Stipe 3–3.5 cm long, 1–1.8 mm thick, straight and brittle, with white pruinescence on upper portion, glabrous on lower portion, white, fibrillose and discoid at base,. Basidiospores 7.5–9.5 × 3.5–4.5 μm, long ellipsoid or lenticular, smooth, hyaline, amyloid.

On decayed wood such as *Pinus*, dead pine needles, dead grass and straw. Distributed in Northeast China.

1371 *Mycena viridimarginata* P. Karst.

Pileus 1–3.5 cm broad, conical or campanulate when young, nearly plane when mature, with a small umbo at centre, glabrous, deeply sulcate, translucently striate at margin, olivaceous brown to greenish brown. Context thin, yellow white to concolourous with pileus. Lamellae decurrent to subadnate, white, yellow white to greyish white, with olivaceous to greenish grey tinge near pileus margin. Stipe 4–7 cm long, 1–3 mm thick, cylindrical, hollow, brittle, with white pruinescence to glabrous, yellow to yellow brown, straight or sometimes slightly curved, black or blackish brown and with white mycelium at base. Basidiospores 6–7.5 × 9–10 μm, broadly elliptical to subglobose, conspicuously apiculate, smooth, hyaline, strongly amyloid.

In summer to autumn, on rotten wood of conifers trees or on trunk or dead branches of fallen wood. Distributed in Northeast China.

1372 *Mycetinis scorodonius* (Fr.) A.W. Wilson & Desjardin

Pileus 1–2.5 cm broad, hemispherical when young, gradually plane when mature, slightly incurved at margin, radially rugulose, dry, glabrous, yellow brown to reddish, gradually becoming lighter. Context thin, whitish to yellow white. Lamellae adnate, narrow, subdistant, often furcate, light to fleshy pink. Stipe 2–6 cm long, 1–2 mm thick, cylindrical or flattened, concolourous with pileus or slightly lighter at apex, dark brown on rest portions. Odour obviously garlic. Basidiospores 7–9 × 3–5 μm, long ellipsoid, smooth, hyaline, amyloid.

In summer, on humus or plant residues in coniferous forests. Distributed in Northeast China.

1373 *Myxomphalia maura* (Fr.) H.E. Bigelow

Pileus 2–5 cm broad, dark grey to dark brown, with looming radial filamentous patterns, viscid when wet, depressed at centre. Context thin, grey white, whitish or translucent. Lamellae slightly decurrent, dirty white to light grey. Stipe 3–5 cm long, 3–5 mm thick, concolourous with pileus, glabrous. Basidiospores 5–5.5 × 3.5–4.5 μm, subglobose to broadly elliptical, thick walled, smooth, hyaline, amyloid.

In summer to autumn, on burnt ground. Distributed in Qinghai-Tibet Regions, China.

1374 *Neolentinus lepideus* (Fr.) Redhead & Ginns

≡ *Lentinus lepideus* (Fr.) Fr.

Pileus 5–16 cm broad, hemispherical or oblate hemispherical, gradually plane or depressed at centre, creamy white to pale yellow brown or pale yellow, with darker or lighter large scales; margin blunt, sometimes lacerated or undulate. Context white to cream, corky when dried. Lamellae white to cream, yellow brown when dried, adnate to decurrent, wide, subdistant, unequal, serrate at edge. Stipe 4–7 cm long, 0.8–3 cm thick, eccentric, subcylindrical, with a tomentose membrane, cream to pale yellow on upper portion, light brown at base, with brown to dark brown scales. Basidiospores 9–13 × 3.5–5.5 μm, subcylindrical, thin walled.

In summer to autumn, on rotten wood in conifers forests, subcaespitose. Edible, it is better to eat when young. Distributed widely in China.

1375 *Omphalina lilaceorosea* Svrček & Kubička

Pileus 1–2.5 cm broad, convex to nearly plane, depressed at centre, umbilicate, ochraceous brown, hygrophanous, plane to slightly incurved at margin, light red brown, lacerated to valvate and black when mature or at age. Context pale pinkish brown to pale brown. Lamellae decurrent, close, pinkish brown to red brown, often with darkish macules. Stipe 1–3 cm long, 5–8 mm thick, cylindrical, often eccentric, pinkish brown to red brown, often with white fibrils. Basidiospores 7.5–8.5 × 4.5–5 μm, elliptical, smooth, hyaline, inamyloid.

On rotten wood in broad-leaved trees. Distributed in Northeast China.

1376 *Omphalotus flagelliformis* Zhu L. Yang & B. Feng

Pileus 4–8 cm broad, infundibuliform when mature, sometimes with a small umbo at centre, red brown to brown. Context orange yellow, with an inconspicuous fishy odour. Lamellae pale orange red to orange yellow. Stipe 5–12 cm long, 1–2.5 cm thick, pale orange red to orange yellow. Basidiospores 4–5.5 × 3.5–4.5 μm, globose, subglobose to broadly elliptical, smooth, subhyaline, inamyloid. Cheilocystidia 10–25 × 5–7 μm, clavate to subfusiform, flagellous at apex.

On humus. Poisonous. Distributed in Central China.

1377 *Omphalotus japonicus* (Kawam.) Kirchm. & O.K. Mill.

Pileus 6–23 cm broad, spherical at first, then expanding to flabelliform, reniform or semicircular, slightly incurved at margin, orange yellow, cinnamon, with scattered scales at centre, dark violaceous at disc, alternate with flocculence to form irregular patterns, cracked. Context pale yellow, with an unpleasant odour when fresh. Lamellae 2 cm broad, decurrent. Stipe lateral, 1.5–2 cm long, 2.5–4 cm thick, brittle, fibrous, with black spots at base of section when cutting, curved. Basidiospores 13–18 μm in diameter, globose, smooth, hyaline.

Superposed on dead trunks of *Fagus* or *Quercus*. Highly poisonous. Distributed in Northeast China.

1378 *Ossicaulis lignatilis* (Pers.) Redhead & Ginns

Pileus 1–3 cm broad, oblate hemispherical at first, then gradually plane to flabelliform, gradually depressed at centre, with white tomentum at depressed portion, glabrous elsewhere, grey to white, humid, incurved to everted at margin, sometimes lacerated to valvate. Context white. Lamellae decurrent, often intertwined at stipe, close, unequal, white, even at edge. Stipe 2–4.5 cm long, 1.5–2.5 mm thick, subcylindrical, eccentric, white, often curved, solid to soft-hollow insides. Basidiospores 5–5.5 × 2.5–3.5 μm, elliptical to oval, smooth, hyaline, inamyloid.

In summer to autumn, gregarious to caespitose on rotten wood of broad-leaved trees in mixed forests. Distributed in Northeast, Central, and Southern China.

1379 *Oudemansiella bii* Zhu L. Yang & Li F. Zhang

Pileus 4–8 cm broad, oblate hemispherical at first, then plane, gradually depressed at centre, glabrous, slight rugose, slightly viscid when wet, yellow brown, grey yellow brown to grey brown, sometimes lighter to grey yellow. Context white, unchanging when injured. Lamellae white, subdistant. Stipe and underground radicant pseudorhiza 13–20 cm long, 0.5–1 cm thick, 5–10 cm long above ground, cylindrical, thickest at ground, whitish at apex, with squamules concolourous with pileus downwards; often with a pseudorhiza up to 10 cm long underground.. Basidiospores 12–16 × 10–13 μm, broadly elliptical, smooth, hyaline, inamyloid.

In summer to autumn, scattered or gregarious on ground in mixed forests. Distributed in Southern China.

1380 *Oudemansiella brunneomarginata* Lj.N. Vassiljeva

Pileus 3–10 cm broad, flat to plane, usually uneven; glutinous when wet, brown, grey brown to grey, sometimes purplish brown, darker at disc. Context white, not special odour. Lamellae subadnate to sinuate, thick and distant, brown, purplish brown to almost black at edge. Stipe 4–10 cm long, 3–10 mm thick, subcylindrical or tapering upwards, hollow, dirty white, with brown, grey brown, purplish brown to nearly black squamules, slightly enlarged at base, usually not radicant. Annulus absent. Basidiospores 13–17 × 9.5–12 μm, amygdaliform to oval. Cheilocystidia densely arranged as a sterile zone, containing brown to pale brown intracellular pigments.

In summer to autumn, on rotten wood of temperate broad-leaved trees. Edible. Distributed in Northeast China.

1381 *Oudemansiella mucida* (Schrad.) Höhn.

Pileus 4–12 cm broad, hemispherical at first, gradually plane, hygrophanous, viscid or glutinous, white, distantly and inconspicuously striate at margin. Context thin, white, soft. Lamellae adnate to sinuate, wide, distant, unequal, white or pinkish. Stipe 5–8 cm long, 0.3–1 cm thick, cylindrical or enlarged at base, fibrous, solid, white on upper portion, slightly with greyish brown tinge in lower portion. Annulus superior, white, membranous. Basidiospores 15.8–23.8 × 14.9–19.5 μm, subglobose, smooth, hyaline.

In summer to autumn in Northern China, but in spring and winter in Southern China, gregarious, subcaespitose or solitary on tree stumps, fallen wood, or rotten wood. Edible. Distributed in Northeast and Northern China.

1382 *Oudemansiella orientalis* Zhu L. Yang

Pileus 1.5–8 cm broad, dirty white to grey brown, glutinous. Context translucent to whitish. Lamellae thick and distant. Stipe 2–8 cm long, 2–8 mm thick, cylindrical, pale brown to whitish, tomentose, enlarged at base, not radicant. Annulus absent. Basidiospores 11.5–14.5 × 9.5–11 μm, broadly elliptical to elliptical, sometimes nearly broadly amygdaliform, smooth, hyaline. Cheilocystidia densely arranged as a sterile zone. Pleurocystidia 120–200 × 28–45 μm, subcapitate at apex.

In summer to autumn, on rotten wood in subtropical forests. Edible. Distributed in Central China.

1383 *Oudemansiella radicata* (Relhan) Singer

≡ *Xerula radicata* (Relhan) Dörfelt

Pileus 3–7 cm broad, pale brown, olivaceous brown to dark brown, glabrous, slimy when wet, hemispherical when young, gradually plane at maturity, with a wide umbo at centre, radially striate. Context thin, fleshy, white. Lamellae sinuate, wide, a little closer, unequal, white. Stipe 6–20 cm long, 0.5–1 cm thick, cylindrical, white at apex, pale brown downwards, subglabrous, longitudinally striate, often spirally so, brittle at pileipellis, fibrous insides, soft, slightly enlarged at base, radicant downwards to form a very long pseudorhiza. Basidiospores 14–18 × 12–15 µm, subglobose to globose, smooth, hyaline.

In early summer to autumn, in broad-leaved forests or at edge of forests. Edible. Distributed in Northeast China.

1384 *Oudemansiella submucida* Corner

Pileus 2–7 cm broad, flat to plane, dirty white, slightly darker at centre, glutinous. Context fleshy, white. Lamellae thick, distant. Stipe 2–8 cm long, 2–8 mm thick, cylindrical, whitish to beige, with white tomentum, enlarged at base, not radicant. Annulus medial and superior, membranous. Basidiospores 18–24 × 16–21 µm, subglobose to broadly elliptical. Cheilocystidia densely arranged as a sterile zone; pleurocystidia 140–210 × 40–50 µm, clavate to fusiform.

In summer to autumn, on rotten wood in subtropical forests. Edible. Distributed in Southern and Central China.

1385 *Oudemansiella yunnanensis* Zhu L. Yang & M. Zang

Pileus 3–7 cm broad, oblate hemispherical to flat, grey, grey brown to yellow brown, sometimes whitish, glutinous Context white. Lamellae thick, distant. Stipe 2–5 cm long, 3–8 mm thick, central to eccentric, white on upper portion, pale brown on lower portion, slightly enlarged at base. Annulus superior, fagacious. Basidiospores 24–38 × 23–33 μm, globose, subglobose, smooth, hyaline.

In summer to autumn, on rotten wood in subtropical alpine and subalpine forests. Edible. Distributed in Qinghai-Tibet Regions, China.

Agarics 6: Species with Names Beginning with P-R

1386 *Panaeolus campanulatus* (L.) Quél.

Pileus 1.5–3 cm broad, conical to campanulate at first, then hemispherical, slightly umbonate at centre, eggshell-coloured to grey brown or with reddish brown tinge, lighter at margin, often lustrous, often cracked at top when dried, changing to grey black in later stage; margin exceeding lamellae, appendiculate with velar remnants. Context thin, grey to slightly brownish. Lamellae adnate or sinuate, close, grey black, with black macules. Stipe 6–9 cm long, 2–4 mm thick, cylindrical, with white pruinescence at first, brown, slightly darker downwards, hollow. Annulus absent. Basidiospores 9.5–13 × 8–9.5 μm, elliptical to nearly hexagonal in frontal view, smooth, dark brown.

In spring to autumn, solitary or gregarious in edge of broad-leaved forests, park lawn, pasture grassland or livestock dung. Poisonous. Distributed in Northeast, Northern, Northwest, Central, and Southern China. Some people believe *P. campanulatus* is a synonym of *P. papilionaceus*.

© Central China Farmer's Publishing House 2025
Y. Li et al., *Atlas of Chinese Macrofungal Resources*, https://doi.org/10.1007/978-981-96-0502-6_7

1387 *Panaeolus cinctulus* (Bolten) Sacc.

Pileus 2–4.5 cm broad, oblate hemispherical to plane, red brown, quickly fading to pale grey brown, hygrophanous, finely rugulose. Context dirty white to pale grey. Lamellae sinuate, pale grey, with dark grey to nearly black spots. Stipe 6–10 cm long, 3–5 mm thick, cylindrical, pale brown and tinged with violaceous, longitudinally finely atriate, covered with dirty white to light grey minute squamules. Annulus absent. Basidiospores 10–12 × 8–9.5 µm, sublimoniform to lenticular, smooth, pale brown to dark brown.

In summer to autumn, on humus in forests. Poisonous. Distributed widely in China.

1388 *Panaeolus cyanescens* (Berk. & Broome) Sacc.

Pileus 1.5–6 cm broad, hemispherical when young, then campanulate to convex, gradually expanded to plane when mature, not viscid; pale brown at first, gradually white or light grey with maturity, occasionally yellow or brown, pale brown to grey brown, easily lacerated in dry conditions, changing to green or blue to bluish black when injured. Context 1–3 mm thick, white, changing to blue to bluish black when injured. Lamellae adnate to subadnate, close, unequal, grey at first, black at maturity, serrate at edge, mottled with spots. Stipe 5–12 cm long, 2–6 mm thick, cylindrical, slightly enlarged at base, white in upper portion, brown or concolourous with pileus in lower portion, changing to bluish black when injured, with white tomentum and striate, hollow. Basidiospores 11–15 × 8–10 µm, limoniform, with an apical germ-pore, smooth, dark brown to smoky black.

Scattered to gregarious on dung piles, humus rich soil in forests or grasslands. Poisonous. Distributed in Central and Southern China.

1389 *Panaeolus fimicola* (Pers.) Gillet

Pileus 1.5–4.0 cm broad, conical to campanulate at first, then plane to oblate hemispherical to hemispherical, obtuse or slightly umbonate at centre, greyish white to grey brown, yellow brown to brown at disc, with dark zones at margin. Context very-thin, greyish white. Lamellae adnate, subdistant, wide, grey brown, gradually changing to black and grey spots, finally black, white at edge. Stipe 2.5–10 cm long, 2–3 mm thick, cylindrical, brown, slightly darker downwards, hollow. Basidiospores 12.5–15 × 8.5–11.5 μm, limoniform, smooth, brown to dark brown.

In summer, on horse dung and surrounding ground. Poisonous. Distributed in Northeast, Northwest, and Inner Mongolia China.

1390 *Panaeolus* cf. *papilionaceus* (Bull.) Quél.

Pileus 2–4 cm broad, campanulate, grey brown to yellow brown, sometimes areolate into scales at centre, sometimes with velar remnants at margin. Context dirty white. Lamellae sinuate, grey black, with darker and lighter macules. Stipe 8–12 cm long, 3–5 mm thick, cylindrical, dirty white to grey brown. Annulus superior, fagacious. Basidiospores 17–22 × 8–12 μm, elliptical, smooth, dark brown, with a germ-pore.

In summer to autumn, on dung or dung piles. Poisonous. Distributed widely in China.

1391 *Panaeolus semiovatus* (Sowerby) S. Lundell & Nannf.

= *Anellaria semiovata* (Sowerby) A. Pearson & Dennis

Pileus 2–5 cm broad, campanulate, dirty white to beige, glabrous to rugulose, viscid when wet, sometimes lacerated into scales at centre. Context dirty white to light grey yellow, unchanging when injured. Lamellae grey brown, mottled with darker spots. Stipe 7–12 cm long, 0.3–0.6 cm thick, cylindrical, concolourous with pileus. Annulus superior to medial, fagacious. Basidiospores 17–20 × 9.5–12 μm, elliptical, smooth, dark brown, with a germ-pore. Cheilocystidia 30–45 × 8–15 μm, lageniform to nearly fusiform.

In summer to autumn, on abandoned pasture or cow and horse dung. Poisonous. Distributed in Northwest and Qinghai-Tibet Regions, China.

1392 *Panellus edulis* Y.C. Dai et al.

Pileus 20 cm broad, hemispherical, flabelliform or reniform, viscid when fresh, covered with a colloidal membrane, yellow, yellow green with brown or violaceous tinge, with short tomentulum, sometimes subglabrous; margin even, incurved at first. Context 1 cm thick, white, soft. Lamellae decurrent, thin, wide, very-close, white to pale yellow. Stipe 1–2 cm long, 1.5–3 cm thick, short, stout, lateral, with spiny hirsute tomentum or scales, pale yellow or yellow green. Taste mild or with a fresh mushroom taste. Basidiospores 4.5–5.5 × 1.2–1.5 μm, smooth, hyaline, botuliform.

In autumn, on dead standing and fallen wood of *Ulmus pumila* and *Betula*. Edible. Distributed in Northeast and Northwest China.

1393 *Panellus stipticus* (Bull.) P. Karst.

Pileus 1–3 cm broad, flabelliform, plane, slightly incurved at margin, semicircular to reniform, irregular at margin, sometimes lacerated or undulate, dry, tomentulose or flocculent, rugose, areolate or furfuraceous when mature, brown to yellow brown, sometimes fading to dirty white. Context fleshy when young, leathery at age. Lamellae adnate, close, often furcate, interveined; white to pale yellow brown, white, pale yellow, slightly brown, pale khaki, orange white or yellow brown to brown, etc. Stipe lateral, short, tapering at base, light cinnamon. Basidiospores 4–6 × 2–2.5 μm, elliptical, smooth, hyaline, amyloid.

In spring to autumn, gregarious on stumps, trunks and dead branches of broad-leaved trees. Poisonous; medicinal. Distributed widely in China.

1394 *Panus ciliatus* (Lév.) T.W. May & A.E. Wood

Pileus 2–6 cm broad, depressed at centre to deeply infundibuliform, not viscid, cinnamon brown to earthy red brown, chestnut brown when dried, sometimes with lavender tinge, thick tomentose, with bristles at margin, concentrically zonate. Context up to 1 mm, white or pale brown, leathery. Lamellae decurrent, close, pale, beige, pale yellow to woody brown, sometimes with lavender tinge. Stipe 2.2–4 cm long, 2.5–8 mm thick, often eccentric, cylindrical, concolourous with pileus, thickly tomentose, with bristles at **junction** with lamellae, fibrous, solid, often with a pseudosclerotia. Basidiospores 5–6.5 × 2.8–3.4 µm, elliptical, smooth, hyaline.

On pseudosclerotia in rotten wood. Distributed in Southern China.

1395 *Panus conchatus* (Bull.) Fr.

Pileus 4–5 cm broad, plane to depressed at centre, cyathiform or conchate; violaceous when young, yellow white, yellow brown or fleshy brown; margin thin, strongly incurved, undulate or shallowly lacerated, tomentose or with a few bristles. Context tough leathery. Lamellae often decurrent, often slightly united at stipe apex, violaceous or lavender, dirty white to pale yellow, pink and even at edge. Stipe 1 cm long, 0.6 cm thick, eccentric to sublateral, short, cylindrical, solid, violaceous at first, then fading to greyish white, short tomentose to with short bristles. Basidiospores 5.4–6.7 × 2.8–3.5 µm, elliptical to short cylindrical, smooth, hyaline.

On rotten wood. Distributed in Northeast, Northern, Northwest, and Central China.

1396 *Panus fulvidus* Bres.

Pileus 1–4 cm broad, hemispherical to convex, sometimes umbilicate at centre, pale yellow brown, yellow brown with a little red to dark brown tinge, dark grey brown at disc, dry, with brown squamules arranged in nearly concentric zones, sometimes with inconspicuous squamules, glabrescent; margin incurved, conspicuously sulcate, even plicate. Context fleshy to leathery. Lamellae sinuate to free, sometimes with a short decurrent tooth, white to pale yellow, slightly narrower, subeven to inconspicuously serrulate at edge. Stipe 2–4 cm long, 0.5–1 cm thick, central to subcentral, cylindrical, subequal, slightly enlarged at base, solid; concolourous with pileus or lighter, sometimes dark brown at base, with a few furfuraceous appendages or furfuraceous squamules. Basidiospores 12–18 × 6–9 µm, subcylindrical, smooth, hyaline.

In summer to autumn, on rotten wood. Distributed in Northeast and Qinghai-Tibet Regions, China.

1397 *Panus giganteus* (Berk.) Corner

≡ *Lentinus giganteus* Berk.

Pileus 6–20 cm broad, oblate hemispherical to suboblate when young, depressed at centre, gradually infundibuliform to phialiform, pale yellow but dark at disc, dry, appendiculate with grey or grey black velar remnants, darker and squamulose at centre; margin strongly incurved and then extended, conspicuously or inconspicuously striate. Context white, with a slight odour. Lamellae decurrent, slightly intertwined, unequal, subclose to close, white to pale yellow, with lamellulae of 3 or 4 lengths. Stipe 5–25 cm long, 0.6–3 cm thick, mostly central, cylindrical, slightly thicker at ground, tapering downwards, with a pseudorhiza about 18 cm long underground, concolourous with pileus, pallid, dirty white to white at apex, tomentose, solid to soft and white insides, base radicant downwards as a pseudorhiza. Basidiospores 6.5–10 × 5.5–7.5 μm, elliptical, smooth, hyaline.

In summer to autumn, solitary or caespitose on rotten wood in evergreen broad-leaved forests. Edible. Cultivable. Distributed in Central and Southern China.

1398 *Panus javanicus* (Lév.) Corner

Pileus 3.5–16 cm broad, depressed at centre to infundibuliform, tomentose or inconspicuous squamulose to glabrous, pale khaki or slightly lighter with age. Context white, tough, leathery. Lamellae decurrent, dirty white or concolourous with pileus, close, narrow, unequal, darker than pileus with age. Stipe 3–18 cm long, 1–3.5 cm thick, cylindrical, eccentric to subcentral, whitish to pale brown, tomentose and squamulose, subcylindrical, solid. Basidiospores 5–6 × 2.5–3.5 μm, elliptical, smooth, hyaline.

Scattered or caespitose on rotten wood. Edible when young. Distributed in Southern China.

1399 *Panus neostrigosus* Drechsler-Santos & Wartchow

=*Panus rudis* Fr.

Pileus 3–10 cm broad, convex, gradually depressed to infundibuliform, pale yellow brown, pale brown at centre, often tinged with violaceous or lavender at margin, densely with long hairs, erect short setaceous hairs or long bristles, with more obviously marginal hairs; incurved at margin, thin, often undulate to slightly lacerated. Context 1.5–2 mm thick at stipe, thinner at margin, leathery, white. Lamellae 1–2 mm broad, decurrent, yellow white to pale yellow brown, or with violaceous tint at edge, close, unequal. Stipe 1–1.8 cm long, 3–9 mm thick, cylindrical or slightly enlarged at base, eccentric to lateral, seldom central, fibrous, solid, concolourous with pileus but normally without violaceous tinge, tomentose to hirsute. Basidiospores 3.5–6 × 1.8–2.8 μm, cylindrical to elliptical, smooth, hyaline.

On rotten wood in mixed and broad-leaved forests. Distributed widely in China.

1400 *Parasola auricoma* (Pat.) Redhead et al.

Pileus 2–4 cm broad, campanulate or hemispherical at first, plane to bamboo-hat shaped when mature, darker to dark brown at centre, becoming lighter to yellow brown outwards, glabrous; margin brown and obviously radially striate, lacerated when mature, gradually deliquescent. Context thin, tinged with pileus colour. Lamellae 1–2 mm wide, free, close, thin, white with slightly brown tinge at first, grey to black at maturity. Stipe 4–6 cm long, 2–3 mm thick, cylindrical, hollow, white to translucent, glabrous, brittle. Basidiospores 10–14 × 6–8 µm, elliptical to oval, smooth, grey black.

In late spring to early summer, scattered on debris and grassland in forests. Distributed in Northeast China.

1401 *Parasola leiocephala* (P.D. Orton) Redhead et al.

≡ *Coprinus leiocephalus* P.D. Orton

Pileus 0.4–1 cm broad, oval to subcylindrical at first, gradually changing to campanulate, then plane, grey yellow to dark yellow, darker at centre, radially striate at margin. Context thin, pale yellow. Lamellae free, grey to grey black, thin. Stipe 5–9 cm long, 1–2 mm thick, cylindrical, white, slender, hollow. Basidiospores 7–8.5 × 4.5–5 µm, elliptical, smooth, blackish brown to black.

Gregarious or solitary on rotten wood in forests. Distributed in Northeast and Northern China.

1402 *Parasola plicatilis* (Curtis) Redhead et al.

Pileus 1–3 cm broad, oval at first, gradually changing to campanulate, then plane, pale grey, slightly depressed at centre, brownish, radially long striate from margin to centre. Context thin, dirty white. Lamellae subfree, distant, grey to grey black, thin, not deliquescent. Stipe 3–7 cm long, 1–2 mm thick, cylindrical, white, glabrous, slender, hollow. Basidia 25–35 × 11–13 µm. Basidiospores 10–12 × 8–10 µm, sublimoniform, smooth, blackish brown to black. Clamp connections present.

Solitary or gregarious on grassland, wood chips or humus in flower nursery. Distributed widely in China.

1403 *Paraxerula ellipsospora* Zhu L. Yang & J. Qin

Pileus 2–5 cm broad, grey brown to light grey, with white to dirty white tomentum. Context very thin, white. Lamellae adnate, distant, white to beige. Stipe 5–7 cm long, 3–5 mm thick, cylindrical, white in upper portion, light grey to pale brown on lower portion, with dirty white tomentum, slightly enlarged at base. Annulus absent. Basidiospores 10–13 × 5.5–7 μm, elliptical to long ellipsoid, smooth, thin walled. Lamellae edge fertile, with scattered cheilocystidia.

On ground in *Pinus* in subtropical mountain, connected with rotten wood underground. Edible. Distributed in Central China.

1404 *Phaeocollybia jennyae* (P. Karst.) Romagn.

Pileus 1.5–4 cm broad, conical to plane or oblate conical, with an umbo; orange brown or waxy brown, with appressed tomentum or glabrous, slightly incurved at margin. Context thin, pale brown. Lamellae close, whitish at first, then rusty, unequal. Stipe 4–5 cm long, 3–4 mm thick, central to eccentric, cylindrical, slightly enlarged at base, tapering and radicant downwards with a pseudorhiza, whitish in upper half portion when young, then gradually changing to brown, glabrous, fibrous, hollow. Basidiospores 4.5–6 × 3–4.5 μm, ovoid, asperulate, not a germ-pore, rust red brown.

In summer to autumn, solitary to scattered on ground in mixed or broad-leaved forest. Distributed in Southern China.

1405 *Phaeocollybia purpurea* T.Z. Wei et al.

Pileus 2–6 cm broad, subconical at first, then plane, with an acute umbo or papilla at centre, violet, grey violaceous to brownish violaceous, glabrous. Context violaceous to purplish grey. Lamellae adnate, violaceous grey to grey brown. Stipe 2–6 cm long, 3–6 mm thick, cylindrical, violaceous grey to grey violaceous, radicant. Basidiospores 3.5–5 × 3–4 µm, oval to sublimoniform, verruculose or asperulate, rusty brown. Cheilocystidia 20–30 × 3.5–5 µm, clavate, apiculate.

In summer to autumn, on ground in subtropical forests. Distributed in Southern and Central China.

1406 *Pholiota adiposa* (Batsch) P. Kumm.

Pileus 5–12 cm broad, oblate hemispherical at first, then plane, slightly prominent at centre, slimy to glutinous when wet, lustrous when dry; lemon yellow, grain yellow, dirty yellow or yellow brown, covered with a layer of transparent mucilage; margin incurved at first, often **appendiculate** with fibrillose velar remnants. Context thick, compact, white to pale yellow, with a mild odour. Lamellae subdecurrent to adnate, close, yellow to rusty yellow. Stipe 4–11 cm long, 0.6–1.3 cm thick, central, slimy, equal or slightly tapering downwards, concolourous with pileus, fibrous. Basidiospores 6–7.5 × 3–4.5 µm, oval to elliptical, thin-walled, smooth, rusty brown.

In spring end to autumn, gregarious and caespitose on fallen trees of broad-leaved trees. Edible. Distributed widely in China.

1407 *Pholiota alnicola* (Fr.) Singer

Pileus 3–7 cm broad, flat hemispherical at first, then plane, slightly prominent at centre, slightly viscid when wet, not hygrophanous, brown or dark cinnamon, with scattered and fagacious scales at margin. Context yellow, unchanging when injured. Lamellae adnate to subsinuate, greyish white or pale yellow at first, then rusty brown. Stipe 6–10 cm long, 0.5–1.1 cm thick, cylindrical, slightly thicker at apex, tapering downwards, yellow brown to dark brown, darker downwards, often curved or twisted, solid at first, hollow with age. Annulus white, fagacious. Basidiospores 8–10.5 × 5–6 μm, oval to elliptical, smooth, yellow brown to rusty brown.

In summer to autumn, gregarious and caespitose on rotten wood in mixed forests. Edible. Distributed in Northeast, Northwest, Central, and Southern China.

1408 *Pholiota aurivella* (Batsch) P. Kumm.

Pileus 5–15 cm broad, oblate hemispherical to convex at first, then plane, obtusely umbonate at centre, viscid when wet, lustrous when dried, golden yellow, rusty yellow in later stage, with appressed subtriangular scales arranged in concentric zones; scales denser at disc, fagacious in later stage; margin incurved at first, **appendiculate** with fibrillose velar remnants. Context pale yellow at first, then lemon yellow. Lamellae adnate or decurrent, close, creamy yellow at first, gradually changing to yellow rust, brown in later stage. Stipe 6–12 cm long, 0.6–1.4 cm thick, cylindrical, often radicant at base, viscid, yellow in upper portion, rusty brown in lower portion, with imbricate reversed scales below annulus which are evanescent in later stage, sometimes curved, solid. Annulus superior, cortinoid, fagacious. Basidiospores are 7–10 × 4.5–6.5 μm, elliptical, smooth, rusty brown.

In autumn, gregarious on rotten wood in forests. Edible. Distributed in Northeast, Northern, and Central China.

1409 *Pholiota flammans* (Batsch) P. Kumm.

Pileus 3–8 cm broad, oblate hemispherical at first, then plane, slightly prominent at centre. Pileus viscid, lemon yellow to orange yellow, with sulphur yellow woolly revolute fibrillose scales arranged in concentric rings, most scales fagacious in later stage; with velar remnants at margin. Context sulphur yellow. Lamellae adnate, sulfur yellow at first, orange brown when injured, rust in later stage. Stipe 4–11 cm long, 5–8 mm thick, central, equal or slightly enlarged at base, sometimes slightly curved, dry, lemon yellow, darker at base, with reversed floccose scales below annulus, solid, becoming hollow at maturity. Annulus superior, yellow, membranous, flocculent, fagacious. Basidiospores 4–5 × 2.5–3 μm, elliptical, smooth, yellow brown.

In summer to autumn, caespitose on fallen coniferous wood, dead standing trees, logs and lumbers. Edible. Distributed in Northeast, Central, and Southern China.

1410 *Pholiota gummosa* (Lasch) Singer

Pileus 2–5 cm broad, hemispherical at first, then convex; papillate at first, gradually obtuse or slightly depressed at centre; initially incurved then plane to irregularly undulate at margin, slightly greenish at first, gradually changing to light lemon colour, light red to light red brown at disc, viscid at first, becoming dry, with pale brown appressed scales which are often **evanescent with age**. Context thin, yellow. Lamellae adnate to subdecurrent, narrow, pale yellow, then changing to pale yellow brown. Stipe 3–8 cm long, 3–7 mm thick, equal, sometimes curved to twisted, pale yellow to lemon colour at first, gradually becoming rusty in lower portion, with patchy veil remnants. Basidiospores 5.5–7 × 3.5–4 μm, elliptical, smooth, yellow brown.

In spring to autumn, on ground or wood in forests. Distributed in Northeast China.

1411 *Pholiota highlandensis* (Peck) Quadr. & Lunghini

Pileus 2–5 cm broad, convex, then gradually plane, viscid, hygrophanous, yellow brown or red brown, lighter at margin. Context thin, yellow to concolourous with pileus, with an inconspicuous smell. Lamellae adnate or sinuate, greyish white to pale yellow at first, gradually changing to cinnamon brown, wide, even or lacerated at edge. Stipe 3–6 cm long, 3–5 mm thick, cylindrical, white to yellow white at apex at first, becoming dirty brown when old, greyish white on lower portion, becoming dark brown when old. Annulus superior, pale yellow to light cinnamon, fibrillose, fagacious. Basidiospores 6–8 × 4–4.5 μm, elliptical to oval, smooth, with an inconspicuous germ-pore, yellow brown.

On burnt ground. Distributed in Central China.

1412 *Pholiota lenta* (Pers.) Singer

Pileus 3–8 cm broad, hemispherical at first, then plane, obtuse at centre, viscid to glutinous, dirty white to yellowish, darker at disc, with white scales at first which are fagacious with age. Context compact, whitish to pale yellow, with a mild smell and taste. Lamellae sinuate to adnate, close, white to pale yellow, then ochraceous cinnamon, white at edge. Stipe 4–8 cm long, 0.4–1.2 cm thick, cylindrical, central, enlarged at base, with pruinose granules above annulus, white or tinged with yellow brown below annulus, covered with white floccose scales, solid to soft insides. Annulus superior, white, large, fagacious. Basidiospores 5–6.7 × 3–4.5 µm, elliptical, smooth, with a tiny germ-pore, yellow brown.

In summer to autumn, gregarious or caespitose on rotten branches or rotten wood in forests. Distributed in Northeast China.

1413 *Pholiota lubrica* (Pers.) Singer

Pileus 4–7 cm broad, oblate hemispherical to hemispherical at first, then plane, prominent at centre, glutinous when wet, red brown at disc, earthy yellow at margin, with yellow gelatinized floccose scales, striate. Context greyish white, yellowish under pileipellis, thick at centre, tough, with mild odour and taste. Lamellae sinuate, adnate to subdecurrent, close, moderately wide, lighter at first, then becoming ochraceous. Stipe 5–8 cm long, 4–6 mm thick, cylindrical, equal or slightly tapering upwards, slightly bulbously enlarged at base, greyish white at first, then becoming brown on lower portion, dry, fibrillose, with flocculence at base, fibrous, solid. Basidiospores 6–7.5 × 3–4 µm, elliptical, smooth, with a tiny germ-pore, pale yellow brown.

In autumn, gregarious on rotten branches or rotten wood in mixed forests. Distributed in Northeast and Central China.

1414 *Pholiota microspora* (Berk.) Sacc.

= *Pholiota nameko* (T. Ito) S. Ito & S. Imai

Pileus 2–8 cm broad, oblate hemispherical at first, then plane, covered with a layer of glutinous mucilage; glabrous; red brown at first, then yellow brown to pale yellow brown, darker at disc; margin thin, striate, incurved at first, with gelatinous velar remnants. Context thick at centre, thin at margin, tough, pale yellow to yellow at first, then cinnamon. Lamellae adnate, close, wide, undulate at edge, grey at first, then rusty. Stipe 2–8 cm long, 0.3–1.3 cm thick, cylindrical, equal or slightly tapering upwards, fibrillose, concolourous with pileus below annulus, glabrous, viscid, solid or slightly hollow, dirty white with pale brown tinge above annulus. Annulus superior, membranous, thin, glutinous, fagacious. Basidiospores 5–6 × 3–4 µm, elliptical to oval, smooth, with a tiny germ-pore, thin-walled, rusty brown.

In summer to autumn, caespitose or gregarious on fallen wood or stumps of broad-leaved trees. Edible. Cultivable, with a trade name of Huagu. Distributed in Northeast and Central China.

1415 *Pholiota populnea* (Pers.) Kuyper & Tjall.-Beuk.

= *Pholiota destruens* (Brond.) Gillet

Pileus 6–15 cm broad, oblate hemisphericalflat hemispherical at first, then plane, viscid when wet, greyish white to whitish or pale yellow to ochraceous, with white scales which are fagacious in later stage; margin incurved at first, plane when mature. Context thick, tough, white, with an inconspicuous smell. Lamellae sinuate to adnate, close to very close, white at first, gradually changing to cinnamon, dark brown at maturity. Stipe 5–10 cm long, 1–2.5 cm thick, central to slightly eccentric, cylindrical, tapering upwards, enlarged at base. Annulus superior, white, wooly, fagacious, then often leaving annular traces on stipe. Basidiospores 7–8.5 × 4–5.5 µm, elliptical to oval, with an obvious germ-pore, slightly truncated at apex, smooth, rusty yellow.

In summer to autumn, solitary or gregarious on trunk or base of broad-leaved forests. Edible. Distributed in Northeast, Inner Mongolia, Northwest, and Qinghai-Tibet, China.

1416 *Pholiota pseudosiparia* A.H. Sm. & Hesler

Pileus 4–9 cm broad, convex, gradually plane to campanulate, dry, with granular scales formed by epidermal cells, dirty ochraceous to dark yellow brown, fading to pale yellow or cinnamon yellow. Context thin, pale yellow, tasteless. Lamellae adnate, pale yellow at first, then gradually becoming ochraceous yellow brown, close, wide, crenulated at margin. Stipe 1.1–2 cm long, 0.5–2 mm thick, cylindrical, equal, light ochraceous in upper portion, pruinose, dirty dark brown in lower portion, flocculent. Basidiospores 6.5–8 × 4–5 µm, oval, long ellipsoid to suboval in frontal view, phaseoliform in side view, smooth, yellow brown.

In summer, scattered on logs or rotten wood in broad-leaved forests. Distributed in Northeast China.

1417 *Pholiota spumosa* (Fr.) Singer

Pileus 2–6 cm broad, oblate hemispherical at first, then plane to depressed at centre, viscid when wet, yellow to sulphur yellow, often greenish, yellow brown at disc, with squamules from ruptured pileipellis in later stage; margin incurved at first, appendiculate with yellow and fagacious floccose cortina, slightly upturned at maturity. Context thin, soft, yellow to sulfur yellow, with mild odour and taste. Lamellae subsinuate to adnate, subclose to close, moderately wide, pale yellow to yellow at first, then changing to brown to rusty brown, often greenish. Stipe 2–10 cm long, 0.2–0.6 cm thick, cylindrical, equal, with white powder and yellow white at apex, brown in lower portion, fibrous, solid insides, then becoming hollow. Basidiospores 6–8 × 3.5–5 µm, oval to elliptical, obtuse at one end, with an obvious but not truncated apical germ-pore, smooth, yellow brown, inamyloid.

In summer to autumn, caespitose or gregarious on ground and fallen wood in coniferous and mixed forests. Distributed in Northeast, Northwest, and Qinghai-Tibet, China.

1418 *Pholiota squarrosa* (Vahl) P. Kumm.

Pileus 3–9 cm broad, campanulate to oblate hemispherical at first, then plane, slightly prominent at centre, incurved at margin, dry, rusty yellow to yellow brown, with red brown revolute hairy scales, often appendiculate with velar remnants at margin when young. Context pale yellow. Lamellae adnate, pale yellow at first, gradually changing to dirty yellow with green or bluish yellow tinge, dirty rust or rusty brown at maturity, even at edge. Stipe 4–12 cm long, 0.4–1.5 cm thick, cylindrical, dry, yellow and smooth above annulus, concolourous with pileus and densely covered with red brown revolute fibrillose scales below annulus. Annulus superior, fibrous, often lacerated, small, dark brown, not fagacious. Basidiospores 6–9 × 4–5 µm, elliptical, smooth, yellow brown.

In summer to autumn, caespitose on fallen wood and stump bases of coniferous and broad-leaved trees. Edible. Distributed in Northeast, Inner Mongolia, Northern, and Central China.

1419 *Pholiota squarrosoides* (Peck) Sacc.

Pileus 3–8 cm broad, oblate hemispherical at first, gradually umbonate-hemispherical, then plane, sometimes depressed at centre, decurved at margin, viscid when wet, light khaki to yellow brown, with cinnamon to chestnut brown erect and pointed scales; scales dry, denser at centre, gradually sparser or absent towards margin; margin incurved at first, often appendiculate with velar remnants. Context thick, white with yellowish tinge. Lamellae adnate, pale yellow at first, then cinnamon, finely serrate. Stipe 5–12 cm long, 0.5–1.2 cm thick, dry, often curved, concolourous with pileus, white and subglabrous above annulus, with chestnut brown or light umber floccose scales below annulus. Annulus superior, pale brown, woolly-membranous, fagacious. Basidiospores 4–5 × 2–3.5 μm, elliptical, smooth, yellow brown.

In summer to autumn, scattered or caespitose on rotten wood or stumps of broad-leaved trees. Edible. Distributed in Northeast, Northern, and Central China.

1420 *Pholiota terrestris* Overh.

Pileus 2–8 cm broad, obtuse convex to subplane, prominent at centre, pale yellow brown, brown to nearly brown, with nearly brown fibrillose scales. Context thick, bright yellow brown to brown, with a mild odour and taste. Lamellae adnate, close, narrow, light at first, then cinnamon to ochraceous brown, uneven at edge. Stipe 3–8 cm long, 4–8 mm thick, cylindrical, equal or tapering downwards, grey, yellow to nearly brown, occasionally discolouring when injured, with white powder above annulus, with dark revolute scales below annulus, solid at first, becoming hollow. Annulus superior. Basidiospores 5–7 × 3.5–5 μm, elliptical to oval, with a germ-pore, smooth, yellow brown, inamyloid.

In spring to autumn, caespitose on ground in forests or roadside. Distributed in Northeast and Central China.

1421 *Pholiota virescens* E.J. Tian & T. Bau

Pileus 2.8–6.4 cm broad, convex, gradually plane, even at margin, slimy, dark vinaceous brown entirely at first, gradually changing to dark green towards to margin in later stage, mostly glabrous, with a few brown scales formed by veils at margin, with brown velar remnants at margin. Context brown, thin, with mild odour and taste. Lamellae adnate, light at first, gradually changing to dark cinnamon brown to dark greenish brown, close, moderately wide, even at edge. Stipe 2–4.3 cm long, 4–5 mm thick, cylindrical, equal, solid, white in upper portion, gradually darker towards base, dirty brown at base, with white to brown flocculent or fibrillose appendages. Annulus superior, fagacious, leaving brown residual marks when fallen off. Basidiospores 6–7.5 × 4–4.5 µm, elliptical, long ellipsoid to oval, smooth, yellow brown.

In summer, scattered on rotten wood of broad-leaved trees. Distributed in Northeast China.

1422 *Physalacria lateriparies* X. He & F.Z. Xue

Pileus 0.5–3 cm broad, 0.5–2.5 mm high, hollow, spherical when young, hemispherical or umbelliform when mature, partially even or rugulosely folked on the downward surface, white. Stipe 1–4.5 mm long, 0.2–0.4 mm thick, thicker near pileus, straightening downwards, tapering to cuneate at base. Column present inside the hollow pileus, short, formed by a bundle of hyphae at junction of stipe and pileus, with radial hyphae from apex. Basidiospores 4.5–5.5 × 2–3 µm, long ellipsoid, smooth, inamyloid.

In summer to autumn, caespitose or gregarious on fallen wood of birch trees in mixed forests. Distributed in Northeast China.

1423 *Pleuroflammula flammea* (Murrill) Singer

Pileus 1–3 cm broad, convex to reniform, fibrillose, changing glabrous when mature, not viscid, non hygrophanous or subhygrophanous, crenulated and with velar remnants at margin, yellow to yellow brown. Context pale yellow, bitter taste. Lamellae free, subclose, moderately wide, cinnamon to brown, then gradually dark brown, light colored at edge. Stipe 4–5 mm long, 1–1.5 mm thick, cylindrical, eccentric, equal to slightly enlarged at base. Light and glabrous above annulus, concolourous with pileus below annulus, tomentose. Annulus superior, membranous or fibrillose, yellow to yellow brown, fagacious. Basidiospores 8–9 × 5–6 µm, oval to elliptical, smooth, yellow brown.

In summer to autumn, gregarious on rotten and dead wood of broad-leaved trees. Distributed in Northeast China.

1424 *Pleuroflammula tuberculosa* (Schaeff.) E. Horak

Pileus 1.5–3 cm broad, convex, incurved at margin, gradually plane, centrally prominent or not, depressed at centre, often with velar remnants at margin, lemon colour at first, becoming orange brown or orange in later stage at centre, with dense appressed squamules. Context yellow. Lamellae adnate to slightly sinuate, greyish white at first, yellow to cinnamon brown in later stage, greyish white and slightly serrulate at edge. Stipe 2–5 cm long, 2–8 mm thick, cylindrical, often curved, equal or slightly enlarged at base, light to lemon yellow in upper portion, deep yellow in lower portion, initially with orange or similarly coloured fibrillose scales similar to those on pileus. Annulus inconspicuous, cortinoid. Basidiospores 6.5–9 × 4–5 μm, elliptical, smooth, yellow brown.

In summer, solitary or caespitose on rotten wood, logs or dead branches of broadleaved trees. Distributed in Northeast, Central and Southern China.

1425 *Pleurotus calyptratus* (Lindblad.) Sacc.

Fruiting bodies typically pleurotoid, fleshy. Pileus 4–12 cm broad, semicircular to conchate when young; smoky grey, pale yellow, milky white, pale brown, greyish white when mature, tomentose or glabrous; incurved and thin at margin. Context 1–3 mm thick, rough, white. Lamellae 2–4 mm wide, decurrent, unequal, subclose to close, with lamellulae at edge, white to dirty yellow. Stipe absent or only as a stipitate base. Inner veil membranous, connected with pileus margin and base, intact or lacerated, white, milky white. Basidiospores 9–14 × 3.75–5.5 μm, cylindrical or long ellipsoid, obviously apiculate at apex, smooth, hyaline, inamyloid.

In spring to autumn, superposed on stumps of broadleaved trees. Edible. Distributed in Northeast, Northwest, and Central China.

1426 *Pleurotus citrinopileatus* Singer

Fruiting bodies caespitose or imbricately superposed, often confluent into large clusters. Pileus 1–3 cm broad, oblate hemispherical, eccentrically infundibuliform or flabelliform, infundibuliform, eccentrically oblate hemispherical or oblate hemispherical, glabrous, bright yellow, yellow to honey yellow, thin, tough, undulate or slightly incurved at margin, sometimes lacerated to valvate, fragrant. Lamellae 1–1.5 mm wide, decurrent, white with yellow tinge, close, often sulcate on stipe. Stipe 2–6 cm long, 5–8 mm thick, eccentric to subcentral, subcylindrical, tapering upwards or equal, white with yellow tinge, confluent with other stipes into clusters at base. Basidiospores 7.5–9 × 2.3–2.8 μm, subcylindrical to long ellipsoid, smooth, hyaline, inamyloid.

In summer to autumn, on dead, fallen, stumps and logs of *Ulmus* trees, occasionally on debilitated standing living trees. Edible and medicinal. Distributed in Northeast, Northern, Northwest, Central, and Southern China.

1427 *Pleurotus djamor* (Rumph.) Boedijn

Fruiting bodies imbricately superposed or caespitose, white, pale pink. Pileus 2.5–11 cm broad, spatulate, pulmonary, conchate or flabelliform, glabrous, tomentose at disc when mature; margin incurved at first, shallowly striate, often short lacerated or even, often proliferous with pileoli. Context thick, thin at margin, brittle, gradually hardened. Lamellae decurrent or deeply decurrent, often woven into furcate or reticulate pattern at stipe, close, very-narrow, thin, even or serrulate at edge. Stipe lateral, 2–12 long, 2–5 cm thick, tomentose. Basidiospores 7–10 × 3–4.5 μm, long ellipsoid, smooth, hyaline, inamyloid.

In summer, on dead wood of broad-leaved trees in tropical areas. Edible. Distributed in Southern China.

1428 *Pleurotus eryngii* (DC.) Quél.

Pileus 3–6 cm broad, oblate hemispherical, circular, conchate, slightly depressed, glabrous, pale yellow to dirty white, densely covered with pale brown scales; margin incurved, sometimes lacerated. Context fleshy, hard when dried, white to cream. Lamellae decurrent, not intertwined, pale violaceous, narrow, subclose, with lamellulae. Stipe 10–30 cm long, 4–10 cm thick, eccentric, thick, cylindrical, firm, with dirty white mycelium at base. Basidiospores 7.5–8.8 × 2.5–4 μm, cylindrical, long ellipsoid, smooth, hyaline, inamyloid.

In late spring to late summer, solitary, gregarious or caespitose at base of the stem of Umbelliferae plants such as *Eryngium foetidum*. Edible, widely cultivated. The natural distribution in China is suspected.

1429 *Pleurotus eryngii* var. *tuoliensis* C.J. Mou

Pileus 5–15 cm broad, spathulate, flabelliform; with white butylaceous spots, incurved at margin, prominent at centre, non-viscid, often areolate. Context white, unchanging when injured, thick. Lamellae 1–2 mm wide, close, decurrent, white. Stipe 5–9 cm long, 2–4 cm thick, lateral, white, thicker in upper portion and thinner at base, solid. Basidiospores 10–14 × 5–6 μm, oval and long ellipsoid, smooth, hyaline, inamyloid.

In spring to autumn, parasitic or saprophyte on the roots of *Ferula sinkiangensis* of Apiaceae. Edible. It has been cultivated on a large scale, and its trade name is Bailinggu. Distributed in Northwest China.

This Chinese cultivated species Bailinggu was once misnamed as *Pleurotus nebrodensis*. The type locality of *Pleurotus nebrodensis,* however, is in Nebrodi, Sicilia, Italy; while the type of ***Pleurotus eryngii* var. *tuoliensis*** is in Xinjiang, China. They are similar but different.

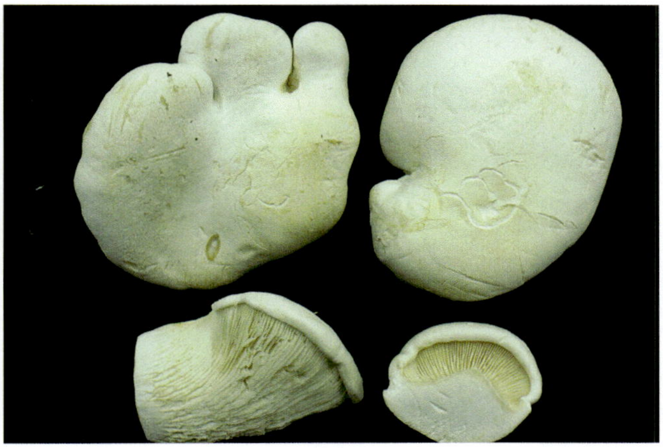

1430 ***Pleurotus nidulans*** **(Pers.) P. Kumm.**

Pileus 2–4 cm broad, oblate hemispherical or flabelliform, reniform, growing radially from a base, depressed to infundibuliform when expanded, yellow brown, hirsute; margin undulate, often incurved, often with brown scales. Context thin, semi-fleshy, not much succulent, 1–2.5 mm thick when dried, leathery, white to pale yellow. Lamellae decurrent to adnate, yellow brown. Stipe absent or rudimentary at base of pileus. Basidiospores 5–6 × 2–2.5 µm, cylindrical to long ellipsoid, smooth, hyaline, inamyloid.

In spring to autumn, on fallen wood and logs of broad-leaved trees. Edible. Distributed in Northeast, Northern, Northwest, Central, and Southern China.

1431 ***Pleurotus ostreatus*** **(Jacq.) P. Kumm.**

Pileus 4–14 cm broad, flat to slightly prominent at first, then plane to flabelliform, reniform, conchate, semicircular, and so on, light grey to dark brown, then gradually changing to dark yellow brown, dry or very viscid when moist, fibrillose-tomentose or glabrous; margin thin, incurved when young, then plane to upturned, sometimes lacerated, estriate. Context thick, fleshy, white, soft when fresh, hardened when dried, strongly renatured when contacted with water. Lamellae 2–4 mm wide, usually decurrent, intertwined on stipe, white, pale yellow to grey yellow. Stipe short or absent, lateral or slightly eccentric if present, 1–3 cm long, 1–2 cm thick, glabrous or densely tomentose, white, solid. Basidiospores 10–11.3 × 3.3–5 µm, cylindrical, long ellipsoid, smooth, hyaline, inamyloid.

In late autumn, on fallen wood, dead trees, stumps, and logs of *Tilia tuan*, *Ulmus pumila*, Acer et al. also on the base of weak living trees. Edible and medicinal. Extensively cultivated. Distributed widely in China.

1432 *Pleurotus porrigens* (Pers.) P. Kumm.

Pileus 1–2.5 cm broad, conchate, semicircular to flabelliform, reniform, growing radially from a base, fleshy, white, subpruinose or flocculent, tomentose at base, undulate and incurved at margin. Context 1–1.5 mm thick, white, fleshy, tasteless. Lamellae 1–2 mm wide, decurrent, radially from a base, unequal, moderately close, subdistant, with lamellulae at margin, white. Stipe absent or as a short stipitate base. Basidiospores 6.3–9.5 × 5.3–7.5 µm, globose, subglobose to elliptical, oval, obviously apiculate, smooth, hyaline, inamyloid.

In summer to autumn, on rotten stumps, fallen wood, rotten wood in mixed or coniferous forests, or at base of *Suaeda glauca*. Edible. Distributed in Northeast and Northwest China.

1433 *Pleurotus pulmonarius* (Fr.) Quél.

Pileus 2.5–10 cm broad or larger, semicircular, flabelliform, reniform, conchate to circular, incurved at margin at first, then plane, depressed at centre or slightly infundibuliform, lacerated into valvate at margin when mature, greyish white or yellow brown, even. Context fleshy, firm, with a strong renaturation, white to creamy white. Lamellae short decurrent, intertwined at stipe, moderately close or subclose, unequal. Stipe absent or present, if present, 0.8–2.5 cm long, 0.7–1.2 cm thick, eccentric to lateral, solid, tomentose at base. Basidiospores 7.5–10 × 3–5 µm, long ellipsoid, cylindrical to elliptical, with an obvious apiculus, smooth, hyaline, inamyloid.

In spring to autumn, on dead wood of broad-leaved trees. Edible. Cultivable. Distributed in Northeast, Central, and Southern China.

1434 *Pluteus admirabilis* (Peck) Peck

Pileus 0.9–3 cm broad, campanulate, hemispherical to convex, slightly umbonate at centre, humid when fresh, bright yellow at first, then light yellow to yellow brown, rugose, striate at margin. Context thin, white to yellow. Lamellae free, unequal, not furcate, close, white at first, yellow to pink in later stage. Stipe 3–5 cm long, 1.5–2.5 mm thick, cylindrical, slightly enlarged at base, whitish, pale yellow to yellow, cartilaginous, soft to hollow insides. Basidiospores 5–6.5 × 4.5–5.5 µm, oval, broadly elliptical to subspherical, smooth, pink.

In summer to autumn, solitary or gregarious on rotten wood of broad-leaved trees. Distributed in Northern and Central China.

1435 *Pluteus atromarginatus* (Konrad) Kühner

Pileus 6–10 cm broad, campanulate to plane, broadly umbonate at centre, rugulose, pale brown, with grey brown to blackish brown fagacious flocculent scales, ridged at margin, often with a narrow blackish brown margin. Context thin, white, brown at pileipellis. Lamellae close, wide, free, unequal, white to light red brown, blackish brown at edge. Stipe 2.5–10 cm long, 2–3 mm thick, concolourous with pileus or lighter, fibrillosely striate, with glandular scales in lower portion, twisted. Basidiospores 4.5–7 × 4–6 μm, broadly elliptical, smooth, pinkish.

In summer to autumn, caespitose or gregarious on conifers and other rotten wood. Edible, but taste not so good. Distributed in Northeast, Northwest, Central, and Southern China.

1436 *Pluteus aurantiorugosus* (Trog) Sacc.

Pileus 3–4 cm broad, oblate hemispherical to flat, orange red to rose red, gradually becoming yellow to orange yellow at margin, with radial short striae at margin. Context thin, white. Lamellae free, close, white at first, then pink, unequal. Stipe 4–5.5 cm long, 3–5 mm thick, cylindrical, yellow to pale yellow, glabrous to squarrulose, with white to pink mycelium at base. Basidiospores 5.5–6.5 × 4–5 μm, broadly elliptical to subspherical, smooth, pink.

Solitary to scattered on rotten wood in forests. Distributed in Central China.

1437 *Pluteus cervinus* (Schaeff.) P. Kumm.

Pileus 4–10 cm broad, hemispherical to convex at first, then gradually plane or flat, viscid at centre, humid, fuliginous brown, dark brown or burnt brown at disc, floccose-tomentose appressedly, undulately lacerated at margin. Context thick, greyish white with reddish tinge. Lamellae close, white at first, then light vinaceous to pinkish brown, free. Stipe 4–11 cm long, 0.5–1.5 cm thick, cylindrical, slightly enlarged to bulbous at base, white, with long dark or blackish brown fibrills, fibrous. Basidiospores 5.5–8 × 4.5–8 μm, subglobose, broadly elliptical or oval, smooth, pink, inamyloid.

Solitary or gregarious on rotten wood of various deciduous trees, but rarely on rotten wood of coniferous trees. Edible, but with poor taste. Distributed in Inner Mongolia, Northeast, Northwest, and Central China.

1438 *Pluteus chrysophaeus* (Schaeff.) Quél.

Pileus 1–6 cm broad, subcampanulate or oblate hemispherical, then flat, slightly prominent at centre, humid, yellow, darkish yellow, tan or brownish yellow, brown to dark brown or wrinkled prominence at apex, even and fine striate at margin. Context thin, brittle, pale yellow white. Lamellae close, slightly thick, wide, yellow brown to fleshy. Stipe 2.5–4.5 cm long, 3–8 mm thick, cylindrical to slightly thicker downwards, slightly enlarged at base, yellow white to white, with longitudinally striate or white fibrillose scales, soft to hollow insides, almost translucent. Basidiospores 6–7.5 × 4.5–6.5 µm, ovoid to subglobose, smooth, pink.

Gregarious on rotten wood in mixed forests. Edibility and toxicity unknown. Distributed in Central China.

1439 *Pluteus diettrichii* Bres.

Pileus 2–5 cm broad, subhemispherical at first, gradually plane to depressed at centre, light grey brown to grey brown, subglabrous or with slightly darker fibrillose scales, often more at centre, dry. Context thin, white. Lamellae white to pale pink, distant, free, unequal. Stipe 3–5 cm long, 2–4 mm thick, subcylindrical, slightly thicker downwards, concolourous with pileus, or whitish, tomentose, brittle, solid to soft insides. Basidiospores 7–9 × 5–6.5 µm, broadly elliptical to subglobose, smooth, pink.

In summer to autumn, on fallen wood or ground in forests. Distributed in Northeast, Northwest, and Qinghai-Tibet Regions, China.

1440 *Pluteus ephebeus* (Fr.) Gillet

Pileus 5–11 cm broad, subhemispherical at first, then gradually plane, grey brown to dark brown, subglabrous or with dark fibrillose scales which are often denser at centre, slightly viscid. Context thin, white. Lamellae subclose, free, unequal, white to pink. Stipe 7–9 cm long, 0.4–1 cm thick, subcylindrical, concolourous with pileus, whitish in upper portion, tomentose, brittle, solid to soft insides. Basidiospores 6.2–8.3 × 4.5–6.2 µm, suboval to elliptical, rarely subglobose, smooth, pink.

In summer to autumn, on fallen wood or ground in forests. Edible, but with poor taste. Distributed in Northern, Northwest, Central, and Southern China.

1441 *Pluteus* cf. *flavofuligineus* G.F. Atk.

Pileus 1.8–2.2 cm broad, nearly bamboo hat-shaped at first, then gradually plane to convex, subglabrous or with brown fibrils, dry, grey yellow to brownish yellow, darker at centre. Context thin, fleshy, white or concolourous with pileus. Lamellae subclose, free, unequal, ventricose, pale yellow. Stipe 3.6–4.2 cm long, 2–4 mm thick, subcylindrical, slightly thicker and with white mycelium at base, longitudinally striate, pale yellow, brittle, solid to soft insides. Basidiospores 5–6 × 4.5–5.5 μm, subglobose to broadly elliptical, smooth, pale pink.

In summer to autumn, on fallen wood or ground in forests. Distributed in Northeast China.

1442 *Pluteus granularis* Peck

Pileus 2.5–6.5 cm broad, oblate hemispherical to plane, slightly umbonate at centre, not viscid, cinnamon to pale grey yellow, with umber or dark brown squamules or minute granules at centre. Context thin, white to greyish white. Lamellae free, unequal, white to pale pink or light cinnamon. Stipe 4–6 cm long, 3–5 mm thick, cylindrical, slightly enlarged at base, with granular or furfuraceous squamules, striate, fibrous, with white tomentum or mycelium at base. Basidiospores 5.5–7.5 × 5–6.5 μm, subglobose or oval, smooth, pale pink. Cheilocystidia 36–60 × 10–25 μm, apiculate or not, thin-walled. Pleurocystidia 58–78 × 20–30 μm, obtuse at apex, thin-walled.

On rotten wood of broad-leaved trees. Distributed in Central China.

1443 ***Pluteus griseodiscus*** **Jiang Xu & T.H. Li**

Pileus7–8 cm broad, subcircular to circular, plane to slightly depressed at centre, white to dirty white, with brown small firils at centre, short striate at margin. Context 6 mm thick at stipe, white. Lamellae free, subventricose, 1.2–1.3 cm broad, subclose to subdistant, unequal, white to pale pink. Stipe 7.5–8 cm long, 5–6 mm thick, subcylindrical, slightly enlarged at base, white, sometimes with translucent longitudinal striae, solid, brittle, with fine white tomentum at base. Basidiospores 6–8 × 5–6 μm, elliptical or subglobose, smooth, pale pink. Cheilocystidia 47–72 × 9–12 μm, clavate, sharp at apex, sometimes irregularly furcate or with apical prongs, thick-walled. Pleurocystidium 65–115 × 12–27 μm, fusiform or nearly lageniform, with 3 ~ 4 prongs or irregular bulges at apex, thick-walled.

On rotten wood of broad-leaved trees. Distributed in Southern China.

1444 ***Pluteus hispidulus*** **(Fr.) Gillet**

Pileus 1–2 cm broad, conical to convex at first, then gradually plane; scales denser at centre, dark, dark brown or olivaceous black; margin grey or pale brown, with greyish white to pink background, covered with fibrils or pale brown squamules. Context 2 mm thick, white to grey, tasteless. Lamellae free, white at first, then pale pink to pink, unequal. Stipe 1.5–3 cm long, 1–2 mm thick, cylindrical, white or pale grey, sericeous, striate, hollow, with white mycelium at base. Basidiospores 5–7 × 4.5–6 μm, subglobose to oval, smooth, pale pink.

In summer to autumn, on rotten wood of broad-leaved trees. Distributed in Central China.

1446 *Pluteus longistriatus* (Peck) Peck

Pileus 2–6 cm broad, hemispherical to plane, brownish grey to dark grey brown at centre, sometimes becoming lighter as grey orange brown, lighter outwards, long striate, sulcate to rugulose, verruculose. Context white. Lamellae free, subclose, widish, unequal, whitish with pink tinge at first, then changing to pink. Stipe 3–7 cm long, 0.4–0.8 cm thick, cylindrical, white to yellow white in upper portion, gradually brown downwards, striate or with fibrillose scales. Basidiospores 6–7 × 5–5.5 µm, broadly elliptical, smooth, pale pink to pinkish yellow.

In autumn, solitary or gregarious on rotten wood. Edible. Distributed in Southern China.

1445 *Pluteus leoninus* (Schaeff.) P. Kumm.

Pileus 2–6 cm broad, subcampanulate or oblate hemispherical at first, then flat, slightly umbonate at centre, glabrous, bright yellow or orange yellow, darker or wrinkled at umbo, striate and sheen at margin, hygrophanous. Context brittle, white with yellow tinge. Lamellae close, slightly wide, white at first, then pink or fleshy. Stipe 3–8 cm long, 0.4–1 cm thick, cylindrical, gradually thicker downwards, slightly enlarged at base, yellow white, fibrous, slightly darker, finely fibrillose-striate, or with dark brown fibrillose squamules, soft to hollow insides. Basidiospores 6–7 × 5–6 µm, subglobose or elliptical or oval, smooth, pale pink to pale pinkish yellow.

In summer to autumn, gregarious or caespitose on fallen rotten wood of broad-leaved trees. Distributed in Northeast, Northwest, and Central China.

1447 *Pluteus nanus* (Pers.) P. Kumm.

Pileus 1.5–3 cm broad, hemispherical, then flat, obtuse or slightly prominent at centre, grey orange brown, dark brown, chestnut brown, fuliginous or umber, even blackish brown at centre, pruinose-tomentulose, often radially rugulose. Context thin, white to concolourous with pileus, not special odour. Lamellae free, close, white at first, then pink or light salmon coloured to pinkish brown, serrulate with white tomentum at edge. Stipe 2–4 cm long, 2–3.5 mm thick, cylindrical, white to whitish, sometimes pale brown, slightly enlarged at base. Basidiospores 6.5–7.5 × 5–6 µm, subglobose to globose, smooth, pink.

In summer to autumn, scattered or gregarious on rotten wood of broad-leaved trees in forests. Distributed in Northeast China.

1448 ***Pluteus pellitus*** (Pers.) P. Kumm.

Pileus 5–7 cm broad, subhemispherical to plane with an umbonate centre, slightly prominent at centre, white, whitish, slightly darker at centre, with pale brown fibrils, sericeous. Context thin, white. Lamellae white to pink, close, wide, free, unequal, serrulate at edge. Stipe 4–7 cm long, 4–6 mm thick, cylindrical, sericeous, white to whitish, slightly enlarged at base and with yellow brown fibrils. Basidiospores 5–8 × 4–6 µm, broadly subelliptical, smooth, pale pink.

In summer to autumn, solitary or gregarious on rotten wood. Edible, but with a poor taste. Distributed in Northeast, Northwest, and Southern China.

1449 *Pluteus petasatus* (Fr.) Gillet

Pileus 4–15 cm broad, campanulate to oblate hemispherical at first, then nearly plane, incurved at margin, slightly umbonate at centre, white, creamy white to nearly greyish white, pale brown to ochraceous yellow at centre, glabrous, with radially grey brown fibrils or squamules, sheeny. Context slightly thicker, soft, white, unchanging when injured. Lamellae close, white at first, then changing to pale pink to pink flesh colour, occasionally flocculent at edge. Stipe 5–15 cm long, 1–1.5 cm thick, cylindrical, stout, slightly enlarged at base, subglabrous, white, solid. Basidiospores 5.5–8 × 3.5–5 μm, broadly elliptical, smooth, pale pink.

In summer to autumn, on rotten wood or sawdusts, often gregarious, subcaespitose, rarely solitary. This species is common on rotten wood or sawdust piles in Changbai Mountains. Edible. Distributed in Northeast and Northern China.

1450 *Pluteus phlebophorus* (Ditmar) P. Kumm.

Pileus 2–5 cm broad, hemispherical to oblate hemispherical or conical at first, then plane, slightly prominent at centre, slightly hygrophanous, orange brown, grey orange brown, light liver brown or yellow brown to dark cinnamon brown or dark fuligineous colour, rugulose or with black ridges, finely cracked when dry, not scales. Context thin, brittle, white, sourish. Lamellae free, subclose, white, then pink to flesh colour. Stipe 2.7–6.5 cm long, 1–7 mm thick, cylindrical, white, often slightly enlarged at base, slightly with sericeous fibrils, solid insides. Basidiospores 6–8 × 4.5–6.5 μm, broadly elliptical or oval, smooth, pink. Cheilocystidia 20–58 × 10–25 μm. Pleurocystidia 35–67 × 13–25 μm, vesicular or with a small apex, thin-walled.

In summer to autumn, solitary, gregarious or caespitose on rotten wood of broad-leaved trees. Distributed in Northeast, Northwest, and Qinghai-Tibet Regions, China.

1451 *Pluteus plautus* (Weinm.) Gillet

Pileus 2.5–3 cm broad, oblate hemispherical, then gradually plane, pinkish grey, pale brown at centre, glabrous at first, then cracked into brown squamules from pileipellis, striate at margin. Context thin, dirty white to nearly yellow or pale brown, not conspicuous odour. Lamellae free, pinkish white to pinkish brown, unequal. Stipe 3–4.5 cm long, 2.5–3.5 mm thick, cylindrical, white, with yellow tinge in lower portions, slightly enlarged at base. Basidiospores 6.5–8 × 5.5–7 μm, broadly elliptical to subglobose or oval, smooth, pale pink. Cheilocystidia 60–80 × 12–30 μm.

In autumn, on rotten wood of coniferous trees. Distributed in Northeast and Qinghai-Tibet Regions, China.

1452 *Pluteus podospileus* Sacc. & Cub.

Pileus 1.5–3.5 cm broad, campanulate at first, then gradually plane, with slightly umbonate at centre, umber, chestnut, grey orange brown, dark red brown or dark brown to nearly blackish brown at centre, soft, tomentose. Context white, beige to concolourous with pileus, not conspicuous odour. Lamellae free, close, unequal, white at first, then changing to cinnamon to pink. Stipe 3–5 cm long, 3–4 mm thick, cylindrical, slightly enlarged or white bulbous at base, white, whitish to pale grey or pinkish, white tomentose, longitudinally brown striate or with squamules at base, soft to hollow insides. Basidiospores 5–7 × 4–5.5 µm, subglobose or broadly elliptical, smooth, pinkish.

In summer to autumn, on rotten wood of coniferous. Distributed in Northeast China.

1453 *Pluteus pouzarianus* Singer

Pileus 3.5–4.5 cm broad, subcampanulate or oblate hemispherical at first, then gradually plane, sometimes slightly convex at centre; grey brown to dark brown, sometimes darker to nearly blackish brown at centre, or sometimes faded to grey brown with pink colour of spores, sometimes very light at margin; subglabrous, sericeous. Context white, thick, odourless. Lamellae close, free, unequal, white at first, then changing to pink. Stipe 5–7 cm long, 6–9 mm thick, cylindrical, white, gradually with pale brown longitudinal striae or with fuliginous brown fibrils from base to top, slightly enlarged with white mycelium at base. Basidiospores 6–8 × 4–5.5 µm, elliptical to oval, smooth, pale pink.

Scattered or fasciculate on rotten wood in coniferous forests. Distributed in Northeast China.

1454 *Pluteus salicinus* (Pers.) P. Kumm.

Pileus 2–5.6 cm broad, hemispherical to oblate hemispherical at first, then plane and slightly umbonate at centre, grey, grey brown to blue grey brown, dark grey at centre, often finely striate. Context whitish, dirty white or greyish. Lamellae close, free, white with pink tinge, unequal. Stipe 5–9 cm long, 3–6 mm thick, cylindrical, often slightly enlarged downwards, white or gradually concolourous with pileus, glabrous. Basidiospores 7–9 × 5–7 µm, elliptical, smooth, pale pink.

In summer to autumn, gregarious on rotten wood. Edible. Distributed in Northeast, Northwest, and Qinghai-Tibet Regions, China.

1455 *Pluteus thomsonii* (Berk. & Broome) Dennis

Pileus 2–3.6 cm broad, umbonate to flat or plane, brown to dark brown, black to grey at centre, maroon to white at margin, radially rugulose to slightly veined, or with reticulate ridges, radially striate at margin. Context thin, white. Lamellae white or grey at first, pink to brown at maturity, gradually close. Stipe 2.4–6 cm long, 1.5–6 mm thick, slightly enlarged at base, lighter than pileus, longitudinally fibrillose striate, attached with small brown powdery granules, hollow, fibrous. Basidiospores 6–8 × 4–6.5 μm, subglobose to broadly elliptical, smooth, wheat straw yellow to pale pink.

In autumn, solitary or gregarious on rotten wood under weeds in broad-leaved forests. Toxicity unknown. Distributed in Northeast, Northwest, Central, and Southern China.

1456 *Pluteus tricuspidatus* Velen.

Pileus 4–11 cm broad, hemispherical at first, oblate hemispherical when expanded, broadly umbonate at centre, yellow brown or dark brown, tomentose, with irregular radial ridges or veins, with dark brown hairy scales or tomentum at margin, with dark brown reticulate ridges at centre, with finer and lighter reticulum at margin, estriate. Context dirty white, thinnish. Lamellae white when young, then changing to pink to pale pinkish brown, free, close. Stipe 5–8 cm long, 0.4–2 cm thick, cylindrical, densely covered with dark brown scales, slightly enlarged at base. Basidiospores 5.5–6.5 × 4–5 μm, subglobose to oval, smooth, pale pink.

In summer to autumn, on rotten wood in coniferous forests. Distributed in Northeast and Southern China.

1457 *Pluteus umbrosus* (Pers.) P. Kumm.

Pileus 3–9 cm broad, campanulate to hemispherical, then expanded to oblate hemispherical, yellow brown to cinnamon, with dark brown hairy scales, obviously with a dark brown reticulate ridges at centre, finely reticulate and lighter in colour at margin, with fine fibrils at margin. Context grey or white. Lamellae whitish when young, changing to pink to pale orange red when mature, dark brown at edge, close, wide. Stipe 3–9 cm long, 0.4–1.2 cm thick, slightly enlarged at base, white, covered with fine brown hairs. Basidiospores 5.3–6.6 × 4.8–5.3 µm, broadly elliptical to subglobose, smooth, pale pink.

In summer to autumn, solitary, gregarious or caespitose on rotten wood of broad-leaved trees including *Populus ussuriensis*. Distributed in Northeast, Northwest, Central, and Southern China.

1458 *Psathyrella bipellis* (Quél.) A.H. Sm.

Pileus 1–4 cm broad, convex, gradually plane, dark violaceous to purplish red at first, then gradually changing to light violaceous to light red brown, wrinkled, conspicuously striate at margin, with velar remnants at margin. Context thin, brittle, light violaceous. Lamellae adnate, close to subdistant, light violaceous at first, gradually changing to dark purplish brown to dark grey, whitish at edge. Stipe 4–12 cm long, 2–5 mm thick, equal, brittle, pale pink to light violaceous in upper portion, whitish in lower portion, glabrous or fibrillose. Annulus absent. Basidiospores 12–15 × 6–8 µm, elliptical, smooth, light purplish brown.

In spring to summer, scattered to gregarious on ground or rotten wood in forests. Distributed in Northeast China.

1459 *Psathyrella candolleana* (Fr.) Maire

Pileus 2–7 cm broad, conical when young, gradually changing to campanulate, plane when mature, appendiculate with lacery velar remnants at margin at first, yellow white, pale yellow to pale brown, translucently striate, lacerated at margin when mature, hygrophanous. Context thin, dirty white to grey brown. Lamellae close, adnate, pale brown to dark purplish brown, serrulate at edge. Stipe 4–7 cm long, 3–5 mm thick, cylindrical, slightly enlarged at base, solid when young, hollow when mature, sericeous, with white fibrils. Basidiospores 6.5–8.2 × 3.5–5.1 µm, elliptical to long ellipsoid, smooth, pale brown.

In summer to autumn, fasciculate on ground in forests, fields, and roadsides, rarely on decaying wooden stumps. Edible. Distributed in Northeast, Northern, Northwest, and Central China.

1460 *Psathyrella gracilis* (Fr.) Quél.

Pileus 2–4 cm broad, hemispherical to oblate hemispherical at first, then gradually plane, yellow to pale brown when young, dark brown to brown when mature, darker at centre, slightly lighter at margin, hygrophanous, radially striate and wrinkled when dried. Context thin, pale brown, odourless. Lamellae close, pale brown to brown, free, unequal, even at edge. Stipe 5–7 cm long, 0.2–0.4 cm thick, cylindrical or slightly enlarged at base, fibrous, brittle, hollow, lighter than pileus, curved when dried. Basidiospores 12–18 × 7–10 µm, long ellipsoid, smooth, with a germ-pore, dark brown.

In spring to summer, on rotten wood in broad-leaved forests or grassland in forests. Distributed in Northeast, Northwest, and Northern China.

1461 *Psathyrella kauffmanii* A.H. Sm.

Pileus 2–6 cm broad, campanulate at first, then gradually plane, with white fibrils when young, then glabrous, translucently striate when wet, hygrophanous at margin, brown to dark greyish brown, with white fibrillose appendage at margin. Context thin, dirty white to grey brown. Lamellae close, adnate, greyish white to pale brown. Stipe 6–7 cm long, 3–5 mm thick, cylindrical, hollow, sericeous, subequal or slightly thinner on upper portion. Basidiospores 7–8 × 4–4.5 µm, long ellipsoid, smooth, dark brown.

In summer, on ground in broad-leaved forests. Distributed in Southern, Central and Northeast China.

1462 *Psathyrella multissima* (S. Imai) Hongo

Pileus 2–5 cm broad, subconical to hemispherical at first, then gradually plane to campanulate, slightly umbonate at centre, yellow brown when wet, hygrophanous, striate at margin. Context thin, pale brown, tasteless. Lamellae close, adnate, pale brown to brown at first, grey black when mature, unequal. Stipe 0.8–2.5 cm long, 3–5 mm thick, cylindrical, tubularly hollow, greyish white, glabrous or with white tomentum. Basidiospores 7–11 × 4.6–5.2 μm, elliptical, smooth, dark red brown to brown.

In summer to autumn, caespitose on ground in broad-leaved forests. Distributed in Northeast and Northwest China.

1463 *Psathyrella obtusata* (Pers.) A.H. Sm.

Pileus 2–5 cm broad, hemispherical to convex when young, then gradually plane, striate at margin, brown to dark brown, hygrophanous. Context thin, dirty white, with mild and light odour. Lamellae adnate, grey brown to dark brown, subdistant, unequal. Stipe 2–8 cm long, 0.5–1.2 cm thick, cylindrical, hollow, subglabrous, white in upper portion, pale brown in lower portion. Basidiospores 6.5–9.5 × 4.4–6 μm, elliptical to long ellipsoid, smooth, pale brown.

In summer, scattered on ground in broad-leaved forests. Distributed in Northeast, Northern, and Southern China.

1464 *Psathyrella olympiana* A.H. Sm.

Pileus 0.5–3 cm broad, conical to convex when young, then gradually plane; with white small fibrils when young, translucently striate when mature, usually undulate at margin, hygrophanous, red brown to dark brown, with white fibrilllose appendage at margin. Context thin, pale brown, tasteless. Lamellae close, adnate, unequal, dirty white to dark red brown. Stipe 3–6 cm long, 3–5 mm thick, cylindrical or slightly enlarged at base, brittle, sericeous, slightly white in upper portion, pale brown at base. Basidiospores 8–9 × 4–5 μm, long ellipsoid, smooth, red brown.

In summer, solitary or scattered on ground in forests. Distributed in Northeast China.

1465 *Psathyrella piluliformis* (Bull.) P.D. Orton

Pileus 2–5 cm broad, hemispherical when young, gradually changing to campanulate to plane, striate at margin, hygrophanous, pale yellow brown at first, then yellow brown, with fibrillose velar remnants at margin. Context thin, with mild and light odour, brown when wet, pale brown when dried. Lamellae close, adnate, grey brown. Stipe 2.5–8 cm long, 3–6 mm thick, cylindrical, slightly enlarged at base, hollow, brittle, ochraceous brown in upper portion, dark brown at base. Basidiospores 4.5–6.5 × 3–4 μm, elliptical to long ellipsoid, smooth, pale brown.

In summer to autumn, fasciculate at base of trees or on ground in broad-leaved forests. Edible. Distributed in Northeast and Central China.

1466 *Psathyrella pygmaea* (Bull.) Singer

Pileus 0.4–2 cm broad, oblate hemispherical at first, then gradually plane, obtuse at centre, translucently striate at margin, hygrophanous, pinkish brown at centre, pale brown at margin. Context thin, grey brown. Lamellae close, slightly white to red brown, adnate, unequal. Stipe 2–3 cm long, 1–2 mm thick, cylindrical, brittle, central, hollow, dirty white at first, gradually changing to pale brown, entirely pruinose. Basidiospores 4.7–6.9 × 3.5–4.9 μm, elliptical to long ellipsoid, smooth, orange brown.

In autumn, gregarious on rotten wood in mixed forests. Distributed in Northeast China.

1467 *Psathyrella spadicea* (P. Kumm.) Singer

Pileus 2–6 cm broad, convex to oblate hemispherical when young, then gradually plane, undulate at margin when mature, pale brown to dark red brown, hygrophanous. Context thin, greyish white, tasteless. Lamellae close, white to red brown, adnate, unequal. Stipe 3–7 cm long, 0.5–1 cm thick, cylindrical, hollow, sericeous, beige white to pale brown, fibrillose, with white mycelium at base. Basidiospores 8–9.5 × 4.5–5.1 μm, elliptical, smooth, pale brown.

In summer, fasciculate on ground in broad-leaved forests. Distributed in Northeast China.

1468 *Psathyrella spadiceogrisea* (Schaeff.) Maire

Pileus 2–5 cm broad, hemispherical to convex at first, then gradually plane, translucently striate at margin, red brown to grey brown, hygrophanous. Context thin, dirty white to pale brown, with a mild taste. Lamellae close, adnate, greyish white at first, gradually changing to pale brown. Stipe 4–7 cm long, 3–5 mm thick, cylindrical, dirty white in upper portion, gradually changing to pale brown downwards. Basidiospores 7.4–9.5 × 4.2–5.5 μm, elliptical to long ellipsoid, smooth, orange brown.

In summer, scattered on ground in broad-leaved forests. Distributed in Southern, Central, and Northeast China.

1469 *Psathyrella spintrigeroides* P.D. Orton

Pileus 1–3 cm broad, hemispherical to convex when young, then gradually plane, with scattered fibrils, with velar remnants at margin, hygrophanous, pale brown to dark brown at centre, lighter at margin. Context thin, dirty white to pale brown, with an inconspicuous odour. Lamellae subclose, adnate, unequal, pale brown to dark brown. Stipe 3–5 cm long, 3–5 mm thick, hollow, cylindrical, dirty white to beige white. Basidiospores 7.7–8.5 × 3.7–5.1 μm, long ellipsoid, smooth, yellow brown to brown.

Solitary on ground in mixed forests. Distributed in Northeast China.

1470 *Psathyrella typhae* (Kalchbr.) A. Pearson & Dennis

Pileus 1.5–2.5 cm broad, hemispherical to convex when young, then gradually plane, with pale brown fibrils at first, then gradually glabrescent, hygrophanous, dark red brown when wet, pale brown when dried, with white fibrils at margin. Context thin, dirty white to pale brown. Lamellae close, adnate, white to pale brown, serrulate at edge. Stipe 2–4 cm long, 2–3 mm thick, cylindrical, central, equal or slightly enlarged at base, hollow, white on upper portion, gradually changing to pale brown to grey brown downwards, with white fibrils entirely. Basidiospores 9–10 × 4.5–5 μm, long ellipsoid, smooth, pale yellow.

On rotten wood in mixed forests. Distributed in Northeast China.

1471 *Psathyrella velutina* (Pers.) Singer

Pileus 4–7 cm broad, campanulate at first, then bamboo-hat shaped, covered with hairy scales, with white velar remnants at first; earthy yellow, earthy brown when young, gradually changing to yellow brown when mature. Context thin, brittle, white. Lamellae free, pale grey to grey black, narrow, unequal. Stipe 4–11 cm long, 5–9 mm thick, cylindrical or slightly enlarged at base, brittle, hollow, with hairy scales in upper portion. Basidiospores 9.2–12.2 × 6.4–7.5 μm, elliptical to long ellipsoid, with obviously verruculum, dark brown.

In spring to summer, gregarious on ground in forests. Edible. Distributed in Northeast China.

1472 *Pseudoarmillariella bacillaris* Zhu L. Yang et al.

Pileus 2–6.5 cm broad, plane, depressed at centre, red brown to brown, densely covered with red brown scales. Context pale yellow, unchanging when injured. Lamellae decurrent, yellow. Stipe 3–7 cm long, 3–6 mm thick, cylindrical, yellow, covered with flocculent scales. Basidiospores 8.5–12 × 2.5–3.5 μm, long ellipsoid to bacilliform, smooth, hyaline. Scales on pileus composed of hyphae.

In summer to autumn, on rotten wood in subalpine forests. Distributed in Qinghai-Tibet Regions, China.

1473 *Pseudoclitocybe cyathiformis* (Bull.) Singer

Pileus 3–5 cm broad, plane, cupulate or shallowly infundibuliform, glabrous, grey to grey brown, hygrophanous, incurved at margin when young, gradually expanded when mature. Context thin, greyish white. Lamellae decurrent, grey or greyish white, close, unequal. Stipe 4–6 cm long, 0.5–1 cm thick, cylindrical, greyish white, with white mycelium at base. Basidiospores 7.5–9.5 × 4.5–6 μm, elliptical, smooth, hyaline.

In summer, scattered on ground or decayed fallen wood in broad-leaved or mixed forests. Distributed in Northeast, Northwest and Qinghai-Tibet Regions, China.

1474 *Psilocybe coprophila* (Bull.) P. Kumm.

Pileus 1–2.5 cm broad, convex to hemispherical, umbilicate at centre when plane, slightly viscid to viscid, glabrous, with fine hairs at margin, grey brown to dark brown, grey yellow brown when dried. Context thin, white, without a special odour. Lamellae adnate, grey brown to dark purplish brown, subdistant, wide, unequal, roughed and granular at edge. Stipe 2–6 cm long, 1–3 mm thick, cylindrical, subequal, yellow brown to grey brown, slightly darker at centre, tomentose in upper portion, dry, hollow. Annulus absent. Basidiospores 10.5–13 × 7.5–8.5 μm, broadly elliptical to nearly hexagonal in frontal view, elliptical in side view, smooth, dark brown.

In summer to autumn, gregarious on dung piles. Poisonous. Distributed widely in China.

1475 *Psilocybe cubensis* (Earle) Singer

Pileus 1.5–2.5 cm broad, conical or hemispherical, then subplane and slightly prominent at centre, with sparse white scales, with white velar remnants at margin, estriate, hygrophanous; yellow when young, then ochraceous or cream, with white tinge from edge when mature, changing to blueish black when injured. Context whitish, changing to blueish black when injured. Lamellae adnate or sinuate, dark grey to dark purplish brown, finally blackish violaceous, white at edge. Stipe 4–12 cm long, 0.4–1.3 cm thick, cylindrical, enlarged at base, glabrous or slightly squamulose below annulus, striate at apex, white to cream or yellow brown, changing to blueish black when injured; soft or hollow insides. Annulus superior, membranous, white, often covered with basidiospores and dark purplish brown. Basidiospores 12–15.5 × 7–9 µm, suboval, smooth, dark brown.

In summer, gregarious or solitary on dung of cattle and horses. Famous poisonous mushroom. Its psilocybin can cause neurotoxicity, rapid toxic reaction, and hallucinogenic effect, which can cause confusion or coma in time and space, and can be fatal when eating a large amount. Distributed in Southern China.

1476 *Psilocybe taiwanensis* Zhu L. Yang & Guzmán

Pileus 2–3 cm broad, subconical to plane, obtuse at centre but with a small papilla, brown to dark brown; lighter at margin, often with white velar remnants. Context white, changing to blue when injured. Lamellae grey brown, dark red brown to dark violet. Stipe 5–8 cm long, 2–5 mm thick, dirty white to pale brown, partially changing to blue when injured. Basidiospores 6–7 × 4–4.5 µm, subelliptical in side view, subrhombic in dorsal view, with a germ-pore, smooth, dark brown.

In summer, fasciculate on humus in forests. Poisonous. Distributed in Southern China.

1477 *Resupinatus applicatus* (Batsch) Grey

Pileus 1–3 cm broad, conchate to pleurotoid, grey brown to blackish brown, glabrous to pruinose at margin, densely tomentose to base, greyish white or pale yellow grey, translucently striate at margin. Context thin, gelatinized, bright brown. Lamellae decurrent, dark brown, narrow and close. Stipe absent, occasionally with pseudostipe, with fine greyish white tomentum at base. Basidiospores 4.5–6 × 4–5 µm, globose or subglobose, hyaline, smooth, inamyloid.

Dorsally to laterally growing on rotten wood or dead branches of deciduous trees. Distributed in Northeast, Central, and Southern China.

1478 *Resupinatus trichotis* (Pers.) Singer

Pileus 0.8–1.5 cm broad, semicircular or reniform, prominent at centre, with thick tomentum, grey to blackish brown or black. Context thin, gelatinized, dark brown. Lamellae radially from a central or eccentric point at base, narrow, moderately distant, brownish grey to grey brown to nearly black. Stipe absent. Attachment base with dark brown or black mycelium. Basidiospores 4–5.5 × 4–4.5 µm, subglobose or globose, smooth, hyaline, inamyloid.

On rotten wood or dead branches of deciduous trees. Distributed in Northeast China.

1479 *Rhodocollybia butyracea* (Bull.) Lennox

Pileus 3–7 cm broad, hemispherical at first, then plane or uplifted, slightly prominent at centre, often hygrophanous, usually dark red brown, brown, khaki or dirty white, darker at centre, lighter to khaki at margin. Context thick at centre, thin at margin, with mild odour. Lamellae adnate to subfree, very close, yellow white to dirty white, unequal, serrulate at edge. Stipe 4–8 cm long, 3–8 mm thick, cylindrical, enlarged at base, pale yellow to khaki, dark brown when dried, with yellow white to pale yellow fine hairs at base, hollow, longitudinally striate. Basidiospores 5–7.5 × 3–4.5 µm, elliptical, smooth, hyaline, inamyloid.

In summer to autumn, solitary or gregarious on ground in coniferous and mixed forests. Edible. Distributed in Northeast, Central, and Qinghai-Tibet Regions, China.

1480 *Rhodocollybia maculata* (Alb. & Schwein.) Singer

Pileus 6–10 cm broad, oblate hemispherical to nearly flat, obtuse or prominent at centre, white or dirty white, often with rusty brown spots or macules, tinged with yellow or brown with age, even and glabrous; margin incurved when young, estriate. Context thick at centre, white, with mild odour or starchy taste. Lamellae adnate or free, white or yellowish, close, narrow, unequal, serrulate at edge, often with reddish-brownish spots. Stipe 5–12 cm long, 0.5–1.2 cm thick, cylindrical, slender, often curved at base, sometimes enlarged at middle and lower portions and radicant at base, longitudinally long striate or with longitudinal twisted grooves, cartilaginous, soft to hollow insides. Basidiospores 6–7.6 × 4–6.5 µm, subglobose, smooth, hyaline, inamyloid.

In summer to autumn, gregarious or subcaespitose on rotten branch layer, rotten wood, or ground in *Pinus* forests. Edible. Distributed in Northwest and Qinghai-Tibet Regions, China.

1481 *Rhodotus asperior* L.P. Tang et al.

Pileus 3–6 cm broad, flat to plane, peach to pink, viscid when wet, sometimes with reticulate wrinkles, incurved at margin. Context pink to whitish. Lamellae subfree, pink to light flesh colour. Stipe 3–5 cm long, 3–8 mm thick, slightly eccentric, cylindrical, whitish to greyish white, slightly enlarged at base. Basidiospores 5–6.5 × 4.5–5.5 µm, broadly elliptical to subglobose, inamyloid, hyaline, thin-walled, densely spinulose-verrucose, with spines 0.5–1.5 µm high and 0.5–1 µm thick.

In summer to autumn, on rotten wood in evergreen broad-leaved forests. Edible. Distributed in Central and Southern China.

1482 *Rhodotus palmatus* (Bull.) Maire

Pileus 1.5–6.2 cm broad, oblate hemispherical to convex, yellow to orange yellow, light red when dried, often with reticulate ridges and irregularly valved. Context elastic, gelatinous. Lamellae subadnate, whitish to pinkish. Stipe 1.5–5.5 cm long, 0.8–1.5 cm thick, often eccentric, pale pink, whitish at base. Basidiospores 5.8–7 × 5.5–6.1 µm, subglobose, verruculose, hyaline, inamyloid.

In summer to autumn, on rotten wood of broad-leaved trees. Edible. Distributed in Northeast China.

1483 *Rickenella fibula* (Bull.) Raithelh.

Pileus 0.3–1 cm broad, oblate hemispherical, umbilicate at centre, viscid, thin, brittle, light yellow or yellow to orange yellow, darker at centre, orange yellow to orange red, reticulate, but indistinctly so when dried. Context white, brittle, tasteless. Lamellae decurrent, distant, unequal, even at edge, white to creamy yellow. Stipe 0.7–5 cm long, 1–2 mm thick, slenderly cylindrical, equal, creamy yellow to pale orange, finely tomentulose. Basidiospores 4–6 × 2–2.5 μm, elliptical, smooth, inamyloid.

In summer to autumn, solitary or scattered on fallen wood and moss. Distributed in Northeast China.

1484 *Ripartites tricholoma* (Alb. & Schwein.) P. Karst.

Pileus 2–5 cm broad, bamboo-hat shaped to gradually plane, depressed at centre, undulate at margin, striate, with fimbriate bristles or tomentum, white, slightly humid to dry. Context fleshy, white. Lamellae adnate, close, wide, white, gradually changing to light cinnamon. Stipe 2.5–5 cm long, 2–5 mm thick, hollow, brittle, white to dirty brown. Spores 4.5–6 × 4–5 μm, subglobose, verruculose, pale yellow.

In autumn, solitary or gregarious on ground in broad-leaved forests. Distributed in Northeast China.

1485 *Roridomyces roridus* (Fr.) Rexer

Pileus 1–3.5 cm broad, hemispherical to campanulate, not viscid, grey brown or dirty yellow, sometimes whitish, obviously striate. Context thin, white. Lamellae adnate or decurrent, white, distant. Stipe 4–8 cm long, 2–5 mm thick, glutinous, greyish white. Basidiospores 8.5–10.5 × 4.5–5.5 μm, elliptical, smooth, hyaline, amyloid.

In summer, caespitose on rotten wood of broad-leaved trees or grassland in mixed forests. Distributed in Northeast China.

1486 *Russula aeruginea* Lindbl.

Pileus 4.5–9.2 cm broad, oblate hemispherical at first, then gradually plane, depressed at centre to shallowly infundibuliform, grey green to dark **aeruginous**, darker at centre, lighter at margin, viscid when wet, radially reticulate at centre; margin striate or estriate, incurved; pileipellis easily peeled off. Context white, thick at centre, thin at margin. Lamellae adnate to adnexed, subclose, mostly equal, white at first, then dirty white to yellow. Stipe 4–8 cm long, 0.8–1.6 cm thick, cylindrical, solid insides when young, hollow and white with age, darker to yellow brown towards to base. Basidiospores 7–8 × 6–7 μm, subglobose to oval, obtuse at apex, verruculose, subhyaline to pale yellow, amyloid.

In summer to autumn, solitary or scattered on ground in broad-leaved or mixed forests. Edible. Distributed in Northeast and Southern China.

1487 *Russula* cf. *albida* Peck

Pileus 2.5–6 cm broad, flat, slightly depressed at centre, white, glabrous; pileipillis viscid easily peeled off; margin even or with inconspicuously short striate ribs at margin. Context brittle, white. Lamellae equal, subclose, adnate to sinuate, white, interveined. Stipe 2.2–6 cm long, 0.5–1.5 cm thick, cylindrical, white, cavernose insides. Basidiospores 8–9 × 7–8 μm, subglobose, elliptical, subhyaline, obtusely spinulose or verrucolose, amyloid.

In summer to autumn, solitary or gregarious on ground in mixed forests. Distributed in Central and Southern China.

1488 *Russula alboareolata* Hongo

Pileus 4.5–8.5 cm broad, flat hemispherical, convex but slightly depressed at centre, complete and incurved at margin when young, plane at margin when mature, white, pinkish white to pinkish fleshy yellow, dirty white to pale yellow at centre, mostly with cyan tinge, viscid when wet, often inconspicuously to slightly obvious cracked, obviously striate. Context white to slightly pink, unchanging when injured. Lamellae distant, adnate, white to pinkish white, equal. Stipe 2.5–4.5 cm long, 0.8–1.5 cm thick, subcylindrical, white. Basidiospores 6–7.5 × 6–7 μm, elliptical to globose, verruculose and weakly incompletely reticulate, subhyaline, amyloid.

Solitary on ground in coniferous, broad-leaved or mixed forests. Distributed in Southern China.

1489 *Russula* cf. *amoena* Quél.

Pileus 5–9 cm broad, hemispherical when young, then oblate hemispherical, then plane, slightly depressed at centre, gradually expanded to plane or slightly reflexed at margin when mature, light purple brown to violaceous, cracked when mature, often inconspicuously striate at margin. Context white, brittle. Lamellae adnate, white or ivory white, close, equal. Stipe 2–5 cm long, 0.8–1.2 cm thick, central, white to light wine red, subcylindrical, hollow, unchanging when injured, slightly thinner at base. Basidiospores 5.5–7.5 × 4–5 μm, subglobose, spinulose subhyaline, amyloid.

Gregarious or scattered on ground in coniferous or mixed forests. Distributed in Central and Southern China.

1490 *Russula atroaeruginea* G.J. Li et al.

Pileus 3–7 cm broad, hemispherical to plane, green to dark green with yellow tinge, darker at centre, lighter at margin. Context white, unchanging, not acrid. Lamellae beige to pale yellow, furcate, with lamellulae. Stipe 4–6 cm long, 1–2 cm thick, whitish. Basidiospores 6.5–8.5 × 6–7.5 μm, subglobose to broadly elliptical, with amyloid ridges and verrucae, inconspicuously amyloid in supraphilar depression.

In summer to autumn, on ground in subalpine dark coniferous forests. Distributed in Qinghai-Tibet Regions, China.

1491 *Russula aurea* Pers.

Pileus 4–9 cm broad, convex or oblate hemispherical at first, then gradually plane to slightly depressed at centre, slightly viscid when wet, orange red or orange yellow, often darker at centre, striate or inconspicuously striate at margin. Context white to pale yellow, fleshy, with mild or slight acrid taste. Lamellae adnate or free, distant to subdistant, wide, ochraceous yellow, yellow at edge, equal, sometimes unequal, interveined, often furcate at stipe. Stipe 3–8 cm long,

1–2.5 cm thick, central, equal, white or cream to golden yellow, fleshy, cavernose insides when young, changing to hollow when mature. Basidiospores 7–9.5 × 6–8.5 μm, oval to subglobose, pale yellow, verruculose or irregularly ridged, connected to an incomplete reticulum, amyloid.

In summer to autumn, solitary or gregarious on ground in mixed forests. Distributed in Northeast and Central China.

1492 *Russula citrina* Gillet

Pileus 3–11.5 cm broad, oblate hemispherical at first, then plane to slightly depressed at centre, honey yellow, ochraceous, yellow to dark yellow, darker at centre, olivaceous green or yellow brown, viscid when wet, estriate and slightly incurved at margin, pileipellis uneasily peeled off. Context white, with acrid odour, with slight acrid taste. Lamellae sinuate to subfree, white to yellow, equal, few furcate, even at edge. Stipe 3.5–8.3 cm long, 1.5–2 cm thick, cylindrical, or slightly enlarged at base, white to pale yellow, darker to yellow or yellow brown at base with age, solid to hollow. Basidiospores 6.5–9.1 × 5.8–8 μm, subglobose, with small spines or tubercular tuberculation, subhyaline, amyloid.

In summer to autumn, scattered or gregarious on ground in mixed forests. Edible. Distributed in Northeast, Northwest, Central, and Southern China.

1493 *Russula cyanoxantha* (Schaeff.) Fr.

Pileus 5–14 cm broad, oblate hemispherical to convex at first, then gradually plane, depressed at centre to infundibuliform, undulate and incurved at margin, variegated, dark violet to dark olivaceous green, then often light blueish brown, greenish grey, often mottled with different colours, slightly viscid when wet or when rain; pileipellis thin, easily peeled off at margin; margin estriate or inconspicuously striate when mature. Context white, pink or light violaceous at pileipellis, with mild odour. Lamellae adnate to slightly decurrent, white, close, unequal, interveined. Stipe 5–10 cm long, 1.5–3 cm thick, fleshy, white, sometimes pink or light violaceous in lower portion, equal, cavernose insides. Basidiospores 7–8.5 × 6.5–7.5 μm, broadly elliptical to subglobose, scatteredly verruculose, some warts connected with lines, hyaline, amyloid.

In summer to autumn, scattered to gregarious on ground in broad-leaved forests. Distributed in Northeast and Central China.

1494 *Russula delica* Fr.

Pileus 3–16 cm broad, convex or oblate hemispherical at first, umbilicate at centre, then gradually plane, depressed at centre to infundibuliform, dirty white, often with ochraceous or brown tinge, sometimes with rusty brown spots, glabrous or tomentulose, not viscid; margin incurved at first, estriate. Context thick, white or whitish, unchanging when injured, with mild to slightly acrid taste, with fruity odour. Lamellae decurrent, white or whitish, subclose, unequal, often with light green tint at edge. Stipe 2–6 cm long, 1.5–4 cm thick, short and thick, solid, equal or tapering downwards, unchanging when injured, glabrous or fibrillose in upper portion. Basidiospores 7.6–9.5 × 7–8.5 μm, oval to subglobose, subhyaline, spinulose to verruculose, slightly reticulated, subhyaline, amyloid.

In summer to autumn, solitary, scattered or gregarious on ground in coniferous or mixed forests. Edible. Distributed in Northeast, Qinghai-Tibet Regions, Inner Mongolia, Central, and Northwest China.

1495 *Russula densifolia* Secr. ex Gillet

Pileus 5.5–10 cm broad, incurved at first, depressed to umbilicate at centre, then changing to infundibuliform, glabrous, dirty white, grey to dark brown. Context thick, white, changing to red to dark brown when injured. Lamellae adnate or decurrent, furcate, unequal, narrow, close, whitish, changing to red brown when injured, dark brown with age. Stipe 2–4 cm long, 1.6–2 cm thick, white, changing to red when injured at first, then dark brown, solid. Basidiospores 7–9.5 × 5.5–7 μm, oval, reticulate, with warts connected with lines, subhyaline, amyloid.

In summer to autumn, solitary or scattered in coniferous, broad-leaved or mixed forests. Poisonous. Distributed widely in China.

1496 *Russula emetica* (Schaeff.) Pers.

Pileus 5–9 cm broad, oblate hemispherical at first, then changing to plane, depressed with age, viscid, glabrous, pale pink to coral red, lighter at margin, ribbed; pileipellis easily peeled off. Context thin, white, red at pileipellis, with bitter taste. Lamellae equal, pure white, distant, sinuate, interveined. Stipe 4–7.5 cm long, 1–2.2 cm thick, cylindrical, white or pink, cavernose insides. Basidiospores 8–10.5 × 7.5–9.5 μm, subglobose, spinulose, hyaline, amyloid.

In summer to autumn, scattered on ground in forests. Poisonous. Distributed in Northeast, Northern, Central, and Southern China.

1497 *Russula flavida* Frost

Pileus 3–7 cm broad, oblate hemispherical at first, depressed at centre, then shallowly infundibuliform, bright golden or ginger yellow, viscid, gradually roughened to powdery, slightly striate to ribbed at margin. Context white, with acrid taste and unpleasant odour. Lamellae adnate to subfree, subclose to distant, dirty white to tinged with pinkish white, equal, interveined and furcate. Stipe 3–7 cm long, 0.8–2.1 cm thick, thick cylindrical, often longitudinally sulcate-lacunular, darker than pileus to golden or dark ginger yellow, cavernose insides. Basidiospores 7.5–9 × 6–7.6 μm, subglobose, spinulose and reticulate, subhyaline to pale yellow, amyloid.

In summer to autumn, solitary or gregarious on ground in mixed forests. Distributed in Southern China.

1498 *Russula foetens* Pers.

Pileus 5–10 cm broad, oblate hemisphericalflat hemispherical at first, then gradually plane, slightly depressed at centre, pale yellow or dirty ochraceous to pale yellow brown, khaki brown at centre, glabrous, viscid, strongly tuberculate-striate at margin. Context thin, dirty white, pale yellow at pileipellis, brittle, with fishy odour; with acrid and bitter taste. Lamellae sinuate, close, wide, dirty white at first, then gradually changing to pale yellow, often with dark spots, generally equal, thick, furcate at base. Stipe 4–10 cm long, 1.5–3 cm thick, stout, equal or slightly tapering downwards, dirty white to dirty brown, often with dark spots with age or when injured, cavernose to hollow insides. Basidiospores 7.5–10 × 7–9.5 μm, globose to subglobose, obviously spinulose or verruculose to ridged, hyaline, amyloid.

In summer to autumn, gregarious or scattered on ground in coniferous or broad-leaved forests. Poisonous. Distributed widely in China.

1499 *Russula fragilis* Fr.

Pileus 1.5–3 cm broad, oblate hemispherical at first, then nearly plane, depressed at centre, dark pink when young, then purplish black at centre, gradually lighter to grey pink to margin, glabrous and sheeny, striate at margin with age; pileipellis easily peeled off. Context white, with fruity fragrance, with acrid and slightly bitter taste. Lamellae sinuate, close, white to creamy white, equal, rarely furcate, sometimes crenulate at edge. Stipe 2.5–6 cm long, 0.5–2.2 cm thick, cylindrical, solid at first, then cavernose to hollow, reticulate, white, changing to yellow with age. Basidiospores 6.3–8.8 × 5.4–7.9 μm, subglobose, verruculose, reticulate and verruculose, subhyaline, amyloid.

In summer, scattered on ground in mixed forests. Poisonous. Distributed in Northeast and Central China.

1500 *Russula griseocarnosa* X.H. Wang et al.

Pileus 9–16 cm broad, flat, often depressed at centre, purplish red to violaceous red brown; pileipellis easily peeled off. Context grey, becoming darker with age, not acrid. Lamellae greyish when mature. Stipe 6–10 cm long, 1.5–3 cm thick. Basidia 40–50 × 10–13 μm. Basidiospores 8–10 × 6.5–8 μm, subglobose to elliptical, with amyloid conical small spines (spines 1.5–2 μm long), subhyaline to pale yellow, amyloid. Cystidia abundant, apiculate at apex, thick-walled at centre.

In summer to autumn, on ground in Southern subtropical evergreen broad-leaved forests. Famous wild edible fungus. Distributed in Southern China.

1501 *Russula guangdongensis* Z.S. Bi & T.H. Li

Pileus 4–7 cm broad, oblate hemispherical-depressed at centre to depressed, orange when young, then orange brown, tinged with hygrophanously grey purplish brown, viscid when wet, obviously tuberculate-striate. Context white to concolourous with pileus, tasteless, with a strong frangipani scent when dried. Lamellae adnate to slightly decurrent, subdistant, about 6 mm wide, mostly equal, creamy white to milky yellow. Stipe 5–11 cm long, 8–12 mm thick, cylindrical, orange yellow to orange brown, tomentulose. Basidiospores 8–11 μm (excluding spines), subglobose, spinulose and incompletely reticulate, pale yellow, amyloid. Pleurocystidia 60–70 × 12–15 μm, subfusiform, slightly thick-walled, yellow.

Scattered to gregarious on ground in broad-leaved forests. Distributed in Southern China.

This species has a strong frangipani scent (fragrant odour of *Plumeria rubra* flowers) and a bright orange stipe, which are unique among the known *Russula* species with strong odour.

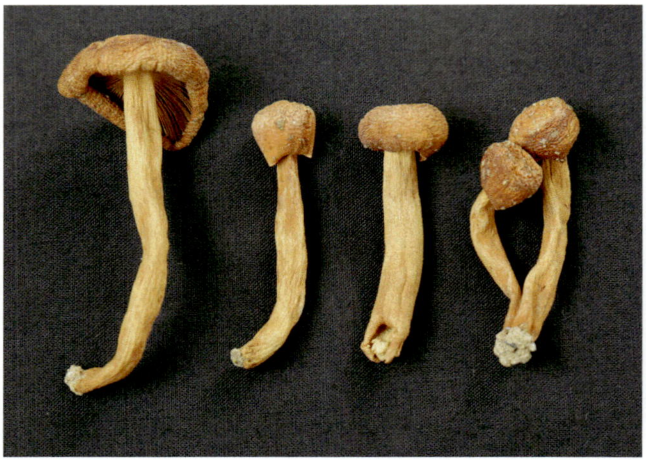

1502 *Russula japonica* Hongo

Pileus 6–15 cm broad, depressed at centre to nearly infundibuliform, slightly incurved at margin, white, often with earthy yellow macules, slightly viscid when wet. Context brittle, white. Lamellae adnate to attached, very close, about 30 per cm at pileus margin, unequal, partially furcate, white, partially changing to creamy yellow to earthy yellow when mature, fragile. Stipe 2.5–5 cm long, 1.2–2.5 cm thick, central to slightly eccentric, white. Basidiospores 6–7 × 5–6 μm, broadly elliptical to subglobose, spinulose, with spines occasionally connected with lines, not reticulate, hyaline, amyloid.

Scattered to gregarious on ground in broad-leaved, mixed or coniferous forests. Poisonous. Distributed in Central and Southern China.

1503 *Russula minutula* var. *minor* Z.S. Bi

Pileus 0.8–2 cm broad, convex at first, then plano-depressed, pink to red, sometimes with purplish red tinge, white to yellow white at margin, viscid when wet, sometimes striate and lacerated. Context 1–1.5 mm thick at stipe, white, brittle. Lamellae 1–3 mm broad, white to yellow white, equal, interveined, adnate to short decurrent. Stipe 0.5–1.5 cm long, 1–3.5 mm thick, cylindrical, white, hollow. Basidiospores 6–8 × 5–7 μm, subglobose, with isolated spines, with some obtuse spines as warts, hyaline to subhyaline, amyloid.

Solitary to scattered on ground in broad-leaved or mixed forests. Distributed in Southern China.

1504 *Russula nigricans* Fr.

Pileus 7–18 cm broad, oblate hemispherical to convex at first, umbilicate at centre, then gradually plane, depressed to infundibuliform, glabrous; dirty white at first, then changing to blackish brown; margin incurved at first, estriate or inconspicuously striate. Context thick, dirty white, changing to light red when injured at first, then changing to black, with fruity odour. Lamellae adnate to sinuate, wide, subdistant to distant, greyish white, unequal, interveined, changing to light red when injured at first, then gradually to grey, finally to black. Stipe 2–7 cm long, 1–2.5 cm thick, equal, glabrous, dirty white at first, then changing to blackish brown, solid insides, brittle, changing to light red when injured at first, then gradually to nearly black. Basidiospores 7–8 × 6–7.5 µm, subglobose, obviously reticulate with connected warts, hyaline, amyloid.

In summer to autumn, gregarious or scattered on ground in broad-leaved or mixed forests. Distributed in Northeast, Central, and Southern China.

1505 *Russula olivacea* (Schaeff.) Fr.

Pileus 6–16 cm broad, plane, slightly depressed at centre, not viscid, variously coloured, olivaceous, light green, sometimes with light purplish red tinge at margin and centre, or uniformly purplish red, violaceous or red brown. Context white or creamy white; odour weak, slightly fruity; taste mild. Lamellae close at first, then gradually distant, furcate, adnate to subfree, beige yellow, gradually changing to ochraceous yellow, unequal, interveined. Stipe 0.4–1.2 cm long, 1–4 mm thick, subcylindrical, solid to cavernose insides, white, usually stained with pink, or pink in upper portion, rarely all pink. Basidiospores 8.5–12 × 7.5–10 µm, subglobose or oval, spinulose, with scattered spines sometimes connected with lines, yellow, amyloid.

In summer to autumn, solitary or gregarious on ground in mixed forests. Edible. Distributed in Northeast, Northern, and Northwest China.

1506 *Russula paludosa* Britzelm.

Pileus 3.5–10.6 cm broad, subhemispherical when young, then gradually plane to depressed at centre, bright red to purplish red, darker at centre, glabrous and sheeny, with minute nodular processes, viscid when wet; margin even and thin, striate with age; pileipellis not separable. Context white; odour mild; taste mild, slightly bitter to slightly astringent. Lamellae adnate to sinuate, white when young, then dirty white or pale yellow, subclose, furcate, sometimes red at edge. Stipe 4.2–9.5 cm long, 1–2.5 cm thick, cylindrical, reticulate, white, red at base, changing to yellow when injured, gradually changing to grey at base. Basidiospores 7.5–10 × 6–8.5 µm, subglobose to suboval, verruculose, subhyaline to pale yellow, amyloid.

In summer to autumn, scattered on ground in coniferous forests. Distributed in Northeast, Qinghai-Tibet Regions, and Central China.

1507 *Russula sanguinea* (Bull.) Fr.

Pileus 3–10 cm broad, convex at first, then gradually plane, depressed at centre to shallowly discoid, bright sanguineous to rose red at first, tinged with violaceous when dried, then often fading irregularly. Context white, unchanging when injured, with fruity fragrance, acrid. Lamellae adnate to slightly decurrent, cream to light ochraceous, subclose, narrow, furcate, equal. Stipe 5–8 cm long, 1–2 cm thick, equal or tapering downwards, usually concolourous with pileus, rarely white, orange yellow when mature or touched, solid insides. Basidiospores 7–8 × 6–7.5 μm, globose to subglobose, hyaline, verruculose, with lines between warts, but not reticulate.

On ground in forests. Distributed in Northeast and Northern China.

1508 *Russula senecis* S. Imai

Pileus 4–10 cm broad, nearly oblate hemispherical to convex at first, then gradually plane, then depressed at centre, revolute at margin, rough, obviously tuberculate-striate; ochraceous yellow brown, dirty yellow to dark yellow brown; slightly viscid. Context pale yellow to dark yellow, with fishy odour, acrid. Lamellae adnate to slightly decurrent, close, dirty white to pale yellow brown, with brown macules at edge, equal or unequal. Stipe 5–9 cm long, 0.4–1 cm thick, equal or tapering downwards, sometimes subfusiform, dirty yellow, dark brown or cinnamon brown, with dark brown glandules, cavernose to hollow insides, brittle. Basidiospores 8–10 × 8–9 μm, subglobose to oval, obviously spinulose, pale yellow, amyloid.

In summer to autumn, solitary or gregarious on ground in mixed forests. Poisonous according to reports. Distributed in Central and Southern China.

1509 *Russula sichuanensis* G.J. Li & H.A. Wen

Pileus 2–5 cm broad, subglobose, hemispherical to campanulate, dirty white, white to pale pink, white at margin, viscid when wet. Context white, unchanging when injured, not acrid. Lamellae yellow to maize yellow, with lamellulae, sometimes, scrobiculate or lacunular. Stipe 3–6 cm long, 0.7–1.5 cm thick, white, subglabrous. Basidiospores 9.5–14 × 8–13 μm, subglobose to globose, with warts and lines connected into reticulate, subhyaline to pale yellow, amyloid.

In summer to autumn, on ground in subalpine dark coniferous forests. Toxicity unknown. Distributed in Qinghai-Tibet Regions, China.

1510 *Russula sororia* Fr.

Pileus 5–9 cm broad, oblate hemispherical to convex at first, then gradually plane, slightly depressed at centre; viscid when wet, glabrous; olivaceous brown to grey brown, grey black at centre, then often fading, lighter at margin; pileipellis easily peeled off at margin, strongly tuberculate-striate at margin. Context white, acrid, with obvious odour. Lamellae free, white at first, then changing to light cream, often with pale brown to light red brown macules, ventricose at centre, close, interveined, unequal. Stipe 3–6 cm long, 1–2 cm thick, equal or tapering downwards, white at first, then changing to pale grey brown, slightly tomentose, solid to hollow insides. Basidiospores 6.5–7.5 × 5.5–6.5 μm, elliptical to subglobose, with small spines or warts, pale yellow, amyloid.

In summer to autumn, solitary or gregarious on ground in forests. Edible. Distributed in Northeast and Qinghai-Tibet Regions, China.

1511 *Russula subnigricans* Hongo

Pileus 5–12 mm broad, oblate hemispherical at first, then nearly plane to depressed or shallowly infundibuliform, pale grey, grey brown to grey blackish brown, sometimes lighter at margin, dry, tomentulose, estriate. Context white, changing to red when injured. Lamellae adnate or subdecurrent, whitish to pale yellow white, changing to red when injured, subdistant, unequal, thick and brittle, not furcate, often interveined. Stipe 3–6 cm long, 1–3 cm thick, cylindrical, grey, lighter than pileus. Basidiospores 6.5–9 × 6–8 μm, subglobose, verrucolose and reticulate, hyaline, amyloid.

In summer to autumn, scattered or gregarious on ground in broad-leaved and mixed forests. Highly poisonous. Distributed in Southern and Central China.

This species is very similar to the non-poisonous *R. nigricans*, but the latter species changing to light red when injured, then obviously changing to black or nearly so.

1512 ***Russula virescens*** (Schaeff.) Fr. s.l.

Pileus 5–12 mm broad, subglobose to convex at first, then gradually plane, slightly depressed at centre; not viscid, or slightly viscid when wet; light green, aeruginous or greyish olivaceous yellow green to grey green; with rusty brown macules, with fine hairs or warts, pileipellis often areolate into small patches or macules, striate at margin when mature, pileipellis inseparable. Context thick, firm and brittle at first, then becoming soft, white, unchanging when injured, or changing to yellow rust when injured, with mild taste and inconspicuous odour. Lamellae free to adnate, white at first, then cream, brown at edge when mature, close, equal, intervenined. Stipe 4–10 cm long, 1–4 cm thick, equal, white, moderately solid or cavernose insides. Basidiospores 7–9 x 6–7.5 µm, subglobose to oval or suboval, verruculose, incompletely reticulate with warts connected with lines, hyaline, amyloid.

In summer to autumn, gregarious on ground in broad-leaved or mixed forests. Famous wild edible fungus. Distributed widely in China.

Agarics 7: Species with Names Beginning with S-X

1513 *Schizophyllum commune* Fr.

Pileus 5–20 mm broad, flabelliform, greyish white to yellow brown, tomentose or hirsute; margin incurved, often valvate, striate. Context 1 mm thick, white, tough, tasteless. Lamellae white to brownish yellow, unequal, longitudinally split down into a deep furrow along the edge. Stipe usually absent. Basidiospores 5–7 × 2–3.5 μm, elliptical or botuliform, smooth, hyaline, inamyloid.

Scattered to gregarious, often superposed on rotten wood or rotten bamboo stems. Edible and medicinal when young; cultivable. Distributed widely in China.

© Central China Farmer's Publishing House 2025
Y. Li et al., *Atlas of Chinese Macrofungal Resources*, https://doi.org/10.1007/978-981-96-0502-6_8

1514 *Simocybe centunculus* (Fr.) P. Karst.

Pileus 1–4 cm broad, flat to plane, chocolate brown, hygrophanous, translucently striate radially at margin. Context thin, light rusty yellow. Lamellae brown to chocolate brown with greyish tinge. Stipe 3–6 cm long, 2–4 mm thick, brown, with white tomentum at apex. Basidiospores 6.5–8.5 × 4.5–5 μm, elliptical to subreniform, with an inconspicuous germ-pore.

In summer to autumn, on rotten wood in coniferous or mixed forests. Distributed in Qinghai-Tibet Regions, China.

1515 *Singerocybe alboinfundibuliformis* (Seok) Zhu L. Yang et al.

Pileus 2–4 cm broad, depressed from centre to base of stipe, white to beige, translucently striate radially at margin. Context thin, white, without special odour. Lamellae decurrent downwards, white, low. Stipe 3–6 cm long, 3–7 mm thick, cylindrical, white to beige, hollow. Basidiospores 6–8 × 4–5 μm, elliptical, smooth, hyaline, inamyloid.

In summer to autumn, on ground or humus in coniferous or mixed forests. Distributed in Central China.

1516 *Singerocybe humilis* (Berk. & Broome) Zhu L. Yang & J. Qin

Pileus 3–5 cm broad, depressed at centre and hollow down to the base of stipe, white to beige, translucently striate radially at margin. Context thin, white, without special odour. Lamellae decurrent, white, low, strongly anastomosing. Stipe 1–2 cm long, 3–5 mm thick, cylindrical, white to beige, hollow. Basidiospores 4.5–6 × 2.5–3.5 μm, elliptical to oval, smooth, hyaline, inamyloid.

In autumn, on ground in tropical to southern subtropical forests. Distributed in Southern China.

1517 *Singerocybe umbilicata* Zhu L. Yang & J. Qin

Pileus 2–4 cm broad, depressed at centre but not hollow down to the base reach to base of stipe, white to beige, undulate at margin. Context thin, white, with a disgusting odour. Lamellae decurrent, white. Stipe 3–5 cm long, 3–6 mm thick, cylindrical, white, beige to pale brown, hollow. Basidiospores 5–8 × 3–4.5 μm, scaphiform, smooth, hyaline, inamyloid.

In summer to autumn, on ground or humus in coniferous or mixed forests. Distributed in Central China.

1518 *Squamanita umbonata* (Sumst.) Bas

Pileus 3.5 cm broad, prominent or umbonate at centre, pale brown, pale grey to dirty white, densely covered with brown, yellow brown to dark brown radially rimose scales. Context white to dirty white. Stipe 3–6.5 cm long, 0.6–1.5 cm thick, cylindrical, dirty white, densely covered with brown, yellow brown to dark brown scales; stipe base bulbous, 3.5–4.5 × 1.5–3.5 cm, pale grey to pale brown. Basidiospores 6–8.5 × 4–5 μm, elliptical, smooth, hyaline, inamyloid.

In summer, on ground in forests. Distributed in Central China.

1519 *Strobilurus* sp.

Pileus 1–3 cm broad, flat to plane, grey to grey brown at centre, whitish at margin. Context white. Lamellae white. Stipe 3–7 cm long, 1–3 mm thick, pale brown to yellow brown, whitish at apex. Basidiospores 3–5 × 2–3 μm, tear-shaped to sunflower seed-shaped, smooth, hyaline. Cystidia apex sublageniform, mostly thin-walled.

In summer to autumn, on the cones of *Pinus armandii* in forests. Edible. Distributed in Central China.

1521 *Stropharia jilinensis* T. Bau & E.J. Tian

Pileus 4–12 cm broad, convex, then gradually plane, undulate at margin, grey violet to dark yellow brown, viscid, with imbricate appressed yellow brown squamules, but with white scales at margin at first. Context thick, greyish white, with mild odour. Lamellae adnate to subdecurrent, pale yellow brown to cinnamon brown, close, moderately wide, serrulate at edge. Stipe 3.5–8 cm long, 0.5–1.5 cm thick, equal to enlarged at base, central, white, with white squamules, hollow, with white flocculent mycelium at base. Annulus membranous to floccose, white. Basidiospores 6–7 × 4–5 μm, elliptical to suboval, smooth, pale grey brown.

In autumn, solitary in coniferous and mixed forests. Edible. Distributed in Northeast China.

1520 *Stropharia aeruginosa* (Curtis) Quél.

Pileus 3–7 cm broad, campanulate to hemispherical, then gradually plane, broadly umbonate at centre, sometimes flat or slightly depressed, covered with a layer of slime at first, with white hairy squamules especially at margin, aeruginous to green, becoming yellow green or grey brownish green with the disappearance of slime layer, usually mottled with irregular yellow macules on aeruginous to green background. Context white. Lamellae adnate to sinuate, greyish white at first, gradually changing to grey purplish brown. Stipe 4.5–7.5 cm long, 4–8 mm thick, equal or tapering upwards, with white rhizomorphs at base. Annulus superior or medial, membranous, fagacious. Basidiospores 8–9.5 × 5–6 μm, elliptical, smooth, pale grey brown.

In summer to autumn, solitary to scattered on rotten wood or debris in coniferous or mixed forests, or on fertile soil. Edible. Distributed widely in China.

1522 *Stropharia rugosoannulata* Farl. ex Murrill

Pileus 5.5–8 cm broad, oblate hemispherical to flat, or convex, slightly viscid when wet, glabrous at pileus margin or covered with floccose scales, with many velar remnants, dark vinaceous brown. Context thick, white. Lamellae adnate, membranous, pale grey to purplish brown; serrulate at edge, white. Stipe 6–12 cm long, 0.9–2 cm thick, cylindrical to subbulbous at base, white to cream. Annulus thick, double layered, striate on upper surface, fagacious at maturity, white. Basidiospores 11–13 × 7.5–8.5 μm, subhexagonal, smooth, pale grey brown.

In summer to autumn, solitary or gregarious in soil or debris. Edible. Distributed widely in China.

1523 *Tectella patellaris* (Fr.) Murrill

Pileus 8–15 mm broad, cylindrical, turbinate to cupulate when young, then gradually expanding to semicircular to conchate, hanging from a dorsal attachment base, incurved and even at margin, slightly viscid, brown, smoky grey to pale pink cinnamon, tomentulose at first, then glabrescent, changing lighter to pale yellow brown to dark brown when mature. Context fleshy or nearly leathery, hardish. Lamellae thin, narrow, yellow brown to pale brown, covered by a membranous veil from pale yellow to fleshy pink. Stipe very short or absent, subcylindrical, curved, eccentric, glabrous, concolourous with pileus. Basidiospores 9.7–11.5 × 3.6–4.2 µm, long cylindrical to botuliform, with a crooked apiculus at base, smooth, hyaline, amyloid.

In summer to autumn, on dead branches in mixed forest of *Betula platyphylla* and *Larix gmelinii*. Distributed in Northeast, Northwest, and Qinghai-Tibet Regions, China.

1524 *Tephrocybe tylicolor* (Fr.) M.M. Moser

Pileus 0.6–1.5 cm broad, convex when young, nearly plane when mature; glabrous, humid, grey brown to dark brown; obviously darker at centre, blackish brown to black; margin lighter, pale yellow to pale brown, striate. Lamellae dirty white, free, distant to moderately close. Stipe 2–5 cm long, 1–3 mm thick, cylindrical, fibrous, sericeous, dirty white to pale brown; curved near base, rooting into substrate, with white to dirty white mycelium at base. Basidiospores 6–7 × 4–5 μm, subelliptical, pyriform or oval, verruculose, obviously apiculate, hyaline.

On decayed debris layer of forests. Distributed in Northeast China.

1525 *Termitomyces aurantiacus* (R. Heim) R. Heim

Pileus 5–15 cm broad, golden yellow, yellow brown to brownish yellow, with an darker obtuse prominence at centre. Context white, unchanging when injured. Lamellae white to pale pink, close. Stipe 5–15 cm long, 0.5–2 cm thick, white to pale yellow, often covered with fibrils or revolute scales. Annulus absent. Pseudorhiza radicant, subcylindrical, white to dirty white. Basidiospores 6–8 × 4–5.5 μm, elliptical, smooth, hyaline, inamyloid.

In summer, on ground in tropical and subtropical zones, connected to underground termite nest. Famous wild edible fungus. Distributed in Southern China.

1526 *Termitomyces bulborhizus* T.Z. Wei et al.

Pileus 10–20 cm broad, oblate hemispherical to plane, pale brown to yellow brown, with an obtuse umbo at centre. Context white, unchanging when injured. Lamellae free, white to pale pink. Stipe 4–10 cm long, 1–6 cm thick; pseudorhiza subcylindrical, white to pale brown; often enlarged to subbulbous at connection of pseudorhiza and stipe, 3–9 cm diam. Basidiospores 6–9 × 4–6 μm, oval to elliptical, smooth, hyaline, inamyloid.

In summer, on ground in subtropical zone, connected to underground termite nest. Famous wild edible fungus. Distributed in Central and Southern China.

1528 *Termitomyces eurrhizus* (Berk.) R. Heim

Pileus 7–12 cm broad, grey to pale grey, conical to plane, acutely umbonate at centre. Lamellae free, white to pinkish. Stipe 5–10 cm long, 0.5–2 cm thick, subcylindrical, white to greyish white. Annulus absent. Pseudorhiza dark brown to blackish. Basidiospores 6.5–8.5 × 4–5 μm, elliptical, inamyloid.

In summer, on ground in tropical and subtropical, connected to underground termite nests. Famous wild edible fungus. Distributed in Southern China.

1527 *Termitomyces clypeatus* R. Heim

Pileus 3–7 cm broad, oblate hemispherical to plane, dark grey, grey to pale grey, with a dark acute umbo at centre. Context white. Lamellae free, white to pale pink. Stipe 4–10 cm long, 0.5–1.5 cm thick, with a subcylindrical pseudorhiza underground, white to dirty white. Basidiospores 5–7.5 × 3.5–4.5 μm, elliptical, smooth, hyaline, inamyloid.

In summer, on ground in tropical, connected to underground termite nests. Famous wild edible fungus. Distributed in Southern China.

1529 *Termitomyces heimii* Natarajan

Pileus 7–15 cm broad, campanulate, bamboo hat-shaped to plane, often umbonate at centre, obtuse at apex, greyish white, pale brownish grey to dark brown at centre, lighter to whitish at margin, glabrous, viscid when wet. Context thick, white, unchanging when injured. Lamellae white to pinkish, free, more or less pinkish and whitish when mature, unequal. Stipe 8–25 cm long, 1–2.5 cm thick, fusiform, white to whitish in upper portion, enlarged near ground, 4–6 cm in diameter at the widest part, changing to dark brown underground, rooting down as a pseudorhiza connected to termite nest underground. Annulus often present at enlarged portion of stipe base, membranous. Basidiospores 7–9 × 4–7 μm, elliptical, smooth, hyaline, inamyloid.

In summer, on ground in tropical zone, connected to termite nests underground. Famous wild edible fungus. Distributed in Southern China.

1530 *Termitomyces microcarpus* (Berk. & Broome) R. Heim

Pileus 1–2.5 cm broad, oblate hemispherical to plane, white to dirty white, with a darker umbo at centre, often reflexed at margin when mature. Context white. Lamellae free, white to pinkish. Stipe 2–5 cm long, 2–4 mm thick. Pseudorhiza subcylindrical, white to dirty white. Basidiospores 6.5–8 × 4.5–5.5 μm, elliptical, smooth, hyaline, inamyloid.

In summer, on termite nests near the ground surface or on ground or roadsides around the termite nests in tropical and subtropical zones. Edible. Distributed in Southern and Central China.

1531 *Termitomyces striatus* (Beeli) R. Heim

Pileus 5–10 cm broad, conical to plane, grey, grey brown to pale brown, acutely umbonate and darker at centre, radially striate, often lacerated at margin. Context white, unchanging when injured. Lamellae free, white to pale pink, close. Stipe 5–15 cm long, 1–2 cm thick; pseudorhiza dirty white. Basidiospores 5.5–7.5 × 3.5–4.5 μm, hyaline, elliptical, smooth, inamyloid.

In summer, on underground termite nests in tropical and subtropical zones. Edible. Distributed in Southern China.

1532 *Tetrapyrgos nigripes* (Fr.) E. Horak

Pileus 5–10 cm broad, flat to plane, pale grey, dark brown to nearly black at centre, depressed, radially sulcate at margin. Context thin, greyish white. Lamellae adnate to subdecurrent, greyish white, subdistant. Stipe 10 mm long, 0.5–1 mm thick, dark grey to black, whitish at apex. Basidiospores 8–9 μm, with 3–5 bulges, mostly with 4 bulges which are up to 7 μm long and 4 μm thick, hyaline, inamyloid.

In summer, on rotten branches and sticks in tropical and subtropical forests. Distributed in Southern China.

1533 *Tricholoma aurantium* (Schaeff.) Ricken

Pileus 5–8 cm broad, oblate hemispherical flat hemispherical to plane, slightly prominent at centre, orange red to red brown, sometimes with greenish to olivaceous tinges, glutinous. Context white, fleshy. Lamellae sinuate, dirty white. Stipe 5–8 cm long, 1–2 cm thick, dirty white, covered with brown to grey scales on middle and lower portions. Basidiospores 5–6.5 × 3.5–4 μm, broadly elliptical to elliptical, smooth, hyaline, inamyloid.

In summer, on ground in coniferous forests. Distributed in Northeast and Qinghai-Tibet Regions, China.

1534 *Tricholoma cingulatum* (Almfelt) Jacobashch

Pileus 3–5 cm broad, oblate hemispherical to plane, slightly prominent at centre, pale grey, whitish at margin. Context white. Lamellae sinuate, white to beige. Stipe 4–7 cm long, 3–8 mm thick, white to dirty white. Annulus medial and superior, white, fagacious, membranous. Basidiospores 4–6 × 2.5–3.5 μm, long ellipsoid, smooth, hyaline, inamyloid.

In summer, on ground in forests. Distributed widely in China.

1535 *Tricholoma fulvum* (DC.) Bigeard & H. Guill.

Pileus 3–7 cm broad, hemispherical at first, then oblate hemispherical to nearly plane, sometimes slightly prominent at centre, brown, darker at centre, viscid when wet, with fibrillose scales, incurved at margin. Context yellow white. Lamellae sinuate, subclose, yellow to dark yellow. Stipe 3–4 cm long, 0.5–1 cm thick, yellow brown, lighter in upper portion, darker downwards, slightly enlarged with white rhizomorphs at base. Basidiospores 6.2–7.7 × 5–5.5 μm, subglobose, smooth, hyaline, inamyloid.

In autumn, solitary or gregarious on ground in forests. Edible. Distributed in Northeast and Central China.

1536 *Tricholoma matsutake* (S. Ito & S. Imai) Singer

Pileus 6–25 cm broad, spherical at first, then oblate hemispherical to plane, slightly prominent at centre, yellow brown to chestnut brown, covered with yellow brown to dark brown fibrillose scales, incurved at margin. Context thick and compact, white, then changing to pale brown, with a strong fragrance. Lamellae sinuate, close, wide, unequal, white, beige to brown. Stipe 10–20 cm long, 1.5–3 cm thick, equal, cylindrical, concolourous with pileus, covered with dark brown to pale brown scales. Annulus superior, fibrillose. Basidiospores 6.5–7.5 × 5.5–6.5 μm, broadly elliptical, smooth, hyaline, inamyloid.

In autumn, solitary to gregarious on ground in pine forests including *Pinus densiflora*, *Pinus thunbergii* and *Pinus sylvestris* var. *mongolica*, often forms fairy rings, forming ectomycorrhiza with *Pinus*. A famous edible fungus, with a trade name matsutake. Distributed in Northeast, Central, and Qinghai-Tibet Regions, China.

1537 *Tricholoma mongolicum* S. Imai

Pileus 5–17 cm broad, hemispherical, plane at maturity, white, glabrous, incurved at margin at first. Context white, thick and compact. Lamellae sinuate, close, unequal, white. Stipe 3.5–6 cm long, 1.5–3.5 cm thick, slightly enlarged at base, white, solid. Basidiospores 6.5–8.1 × 4–5.8 μm, elliptical, smooth, hyaline, inamyloid.

In summer to autumn, on grassland, often forming fairy rings. Famous edible fungus. Distributed in Northeast, Northern, and Inner Mongolia, China.

1538 *Tricholoma myomyces* (Pers.) J.E. Lange = *Tricholoma terreum* (Schaeff.) P. Kumm

Pileus 3–5 cm broad, oblate hemispherical to plane, pale grey, grey to brownish grey, with appressed fibrillose scales. Context fleshy, white. Lamellae sinuate, white to beige. Stipe 3–5 cm long, 0.4–1 cm thick, white to dirty, subglabrous. Basidiospores 5.5–7 × 4–5 μm, elliptical to broadly elliptical, smooth, hyaline, inamyloid.

In summer, on ground in forests. Edible. Distributed widely in China.

1539 *Tricholoma pardinum* (Pers.) Quél.

Pileus 5–10 cm broad, oblate hemispherical to plane, dirty white, covered with brown to dark brown scales. Context fleshy, white. Lamellae sinuate, dirty white to pale brown. Stipe 7–10 cm long, 2–3 cm thick, white, covered with pale brown scales, enlarged at base. Basidiospores 8–10 × 6.5–7.5 μm, elliptical to broadly elliptical, smooth, white, inamyloid. Cheilocystidia larger.

In summer, on ground in forests. Poisonous. Distributed widely in China.

1540 *Tricholoma populinum* J.E. Lange

Pileus 5–14 cm broad, oblate hemispherical to convex at first, then gradually plane; incurved at margin at first, then plane to undulate, viscid, pale brown, red brown at centre, lighter towards to margin, covered with brown squamules. Context thick, dirty white, becoming darker when injured. Lamellae sinuate, close, narrow, dirty white, gradually changing to light red brown, unequal, changing to darker when injured. Stipe 4–6 cm long, 2–3.5 cm thick, thicker, equal to enlarged at base, solid to soft insides, white, changing to light red brown when injured. Basidiospores 4.8–5.5 × 3–5 μm, oval to subglobose, smooth, hyaline, inamyloid.

In autumn, gregarious or scattered on sandy soil in *Populus* forests. Edible. Distributed in Northeast and Northern China.

1541 *Tricholoma psammopus* (Kalchbr.) Quél.

Pileus 3–10 cm broad, convex, then plane, pale yellow brown to rusty brown, darker at centre, viscid. Context thick, white. Lamellae sinuate, close, unequal, white at first, then gradually changing to straw yellow, with rusty spots. Stipe 4–9 cm long, 1–2 cm thick, with small granules on upper portion, with rusty brown fibrillose scales in lower portion. Basidiospores 6–7 × 4–5 μm, elliptical, smooth, hyaline.

In summer to autumn, gregarious or caespitose on ground in coniferous or broad-leaved forests. Distributed in Northwest and Central China.

1542 *Tricholoma saponaceum* (Fr.) P. Kumm.

Pileus 6–10 cm broad, oblate hemispherical to plane, slightly prominent at centre, dark grey brown, olivaceous elsewhere, changing to yellow to dirty white toward margin, not viscid. Lamellae sinuate, beige, distant. Stipe 10–16 cm long, 1–2 cm thick, tapering downwards, white, with white to grey scales, with pink macules at base. Pileus context white, stipe context pale yellow, with soapy smell. Basidiospores 4–5 × 3–3.5 μm, elliptical, smooth, hyaline, inamyloid. Clamp connections present.

In summer, on ground in broad-leaved or mixed forests. Poisonous. Distributed in Northeast and Central China.

1543 *Tricholoma sejunctum* (Sowerby) Quél.

Pileus 3–9 cm broad, subconical to convex at first, then gradually plane, prominent at centre, slightly viscid when wet, yellow or pale yellow green, darker at centre, with dark green fibrils, even or undulate at margin, sometimes with velar remnants. Context slightly thick, white, pale yellow near pileipellis, slightly bitter. Lamellae adnate to sinuate, white to grey pink at first, then pale yellow at pileus margin, close, wide, unequal. Stipe 5–12 cm long, 0.8–2.5 cm thick, equal, sometimes enlarged at base, white in upper portion, gradually changing to pale yellow downwards, sometimes pink at base, solid to soft insides, powdery in upper portion, with fine fibrils in lower portion. Basidiospores 6.5–7.5 × 4.5–6 µm, subglobose to broadly elliptical, smooth, hyaline, inamyloid.

In autumn, gregarious on ground in mixed forests. Distributed in Northeast and Central China.

1544 *Tricholoma* cf. *sulphureum* (Bull.) P. Kumm.

Pileus 4–7 cm broad, oblate hemispherical to plane, slightly prominent at centre, fox brown with typical sulfur yellow tinge, not viscid. Context pale yellow, with a strong coal gas smell. Lamellae sinuate, yellow, close. Stipe 8–10 cm long, 0.6–1 cm thick, yellow, greenish in lower half portion. Basidia 35–45 × 8–9.5 µm. Basidiospores 8.5–10 × 5–6 µm, elliptical to subamygdaliform, smooth, hyaline, inamyloid.

In summer, on ground in mixed forests. Poisonous Distributed in Northeast and Central China.

1545 *Tricholoma vaccinum* (Schaeff.) P. Kumm.

Pileus 3–5 cm broad, oblate hemispherical to plane, light red brown, covered with dark red brown curly scales. Context thick, fleshy. Lamellae sinuate, pale pink. Stipe 3–8 cm long, 0.5–1.5 cm thick, concolourous with pileus. Basidiospores 6.5–7.5 × 5–6 μm, broadly elliptical, smooth, hyaline, inamyloid.

In summer, on ground in coniferous and mixed forests. Distributed in Northeast and Qinghai-Tibet Regions, China.

1546 *Tricholomopsis decora* (Fr.) Singer

Pileus 2–7 cm broad, convex at first, incurved at margin, then gradually plane, depressed at centre, sometimes undulate at margin, yellow, densely covered with pale brown to grey brown squamules, darker at centre. Context pale yellow to yellow, thin. Lamellae adnate to decurrent, yellow, close, unequal. Stipe 2–8 cm long, 5–8 mm thick, equal, pale yellow. Basidiospore 5.5–7.5 × 4–5.5 μm.

In summer to autumn, solitary, scattered or gregarious on rotten wood of conifers. Distributed in Northeast and Central China.

1547 *Tricholomopsis rutilans* (Schaeff.) Singer

Pileus 5–10 cm broad, oblate hemispherical to plane, yellow brown to brownish yellow, darker at centre, densely covered with red brown scales. Context 3 mm thick, yellow to yellow brown. Lamellae pale yellow to yellow. Stipe 5–10 cm long, 0.5–2 cm thick, pale yellow to yellow, covered with red brown scales. Basidiospores 6–7.5 × 4–5.5 μm, elliptical to long ellipsoid, smooth, hyaline, inamyloid. Cheilocystidia 50–120 × 10–25 μm, with coloured contents insides.

In summer, on rotten wood in forests. Distributed widely in China.

1548 *Tricholomopsis sasae* Hongo

Pileus 1–6 cm broad, oblate hemispherical to plane, slightly prominent or lower at centre, brown khaki or khaki to light earthy red or yellow, with floccose squamules especially denser at centre. Context thin, yellow. Lamellae subsinuate, white or yellow, pruinose at edge, close, unequal. Stipe 2–5 cm long, 3–7 mm thick, cylindrical, slightly curved, dirty white to pale yellow, hollow. Basidiospores 5–6.5 × 4–5.3 μm, broadly elliptical or subglobose, smooth, hyaline, inamyloid.

In summer to autumn, on rotten wood, debris in broad-leaved or mixed forests or grasslands. Edible. Distributed in Southern China.

1549 *Tricholosporum porphyrophyllum* (S. Imai) Guzmán

Pileus 5–16 cm broad, dirty brown at first, convex, then gradually plane, slightly depressed at centre when mature, decurved at margin, irregularly cracked, khaki to yellow brown, darker at centre, slightly viscid, lighter at margin. Context white to yellowish white, without special odour. Lamellae close, sinuate, violaceous, then gradually lighter to yellow brown, changing to brown when injured. Stipe 4.5–15 cm long, 0.6–2.5 cm thick, equal, enlarged to bulbous at base, pale yellow to yellow brown, hollow at maturity. Basidiospores 5–8 × 5–7.5 μm, cross-shaped.

In summer to autumn, on ground in forests. Distributed in Northeast China.

1550 *Trogia infundibuliformis* Berk. & Broome

Pileus 2–4 cm broad, infundibuliform, pink to light fleshcolour, sometimes brown. Context thin, white to pinkish, flexible. Lamellae decurrent, narrow, distant, pinkish to dirty white. Stipe 1–3 cm long, 2–4 mm thick, subcylindrical, tough, with white mycelium at base. Basidiospores 7–9 × 3.5–5 μm, elliptical, smooth, hyaline.

In summer to autumn, on rotten wood in tropical to south subtropical forests. Distributed in Southern China.

1551 *Trogia venenata* Zhu L. Yang et al.

Pileus 1–6 cm broad, flabelliform to valvate, pink to light fleshcolour, sometimes dirty white to white. Context thin, white to pinkish, flexible, tasteless. Lamellae decurrent, narrow, distant, pinkish to dirty white. Stipe 0.3–2 cm long, 2–4 mm thick, subcylindrical, tough, with white mycelium at base. Basidiospores 6–8 × 4–5 μm, elliptical to pip-shaped, smooth, hyaline.

In summer to autumn, on rotten wood in subtropical evergreen broad-leaved or mixed forests. Hypertoxic. Distributed in Central and Southern China.

1552 *Tubaria confragosa* (Fr.) Kühner

Pileus 3–6 cm broad, convex to plane, decurved downwards at margin, dry, with floccose or scaly appendages, with velar remnants at margin, red brown or brown when wet, hygrophanous, fading to pale brown or cinnamon at centre at first, then fading to pink to cinnamon elsewhere. Context pink to cinnamon. Lamellae close, pale yellow or brown. Stipe 4–7.5 cm long, 4–6 mm thick, cylindrical, fibrous, red brown or vinaceous red, white and powdery at base, hollow. Annulus superior, white. Basidiospores 6–7.5 × 4.5–5 μm, elliptical to long, smooth, ochraceous to yellow brown or brown to dark brown.

In summer to autumn, gregarious or scattered on rotten wood in coniferous forests. Distributed in Northeast and Central China.

1553 *Tubaria furfuracea* (Pers.) Gillet

Pileus 1–3 cm broad, convex to plane, incurved at margin at first, then expanded to undulate, humid when fresh, yellow brown or pale yellow, densely with white tomentulum, hygrophanously striate at margin. Context thin, yellow to khaki. Lamellae pale yellow to yellow brown. Stipe 2.5–5 cm long, 1.5–4 mm thick, subequal, yellow to yellow brown, hollow, often densely with white flocculent cottony mycelium at base. Annulus inconspicuous, often as a fibrillose zone. Basidiospores 6–9 × 4.5–5 μm, obovate to elliptical, smooth, ochraceous yellow to fleshcolour or pale yellow.

In spring to summer, scattered or gregarious on rotten wood, ground and moss in coniferous forests. Distributed in Northeast, Northwest, and Central China.

1554 *Volvariella bombycina* (Schaeff.) Singer

Pileus 5–13 cm broad, convex to campanulate at first, then gradually plane, white to pale yellow, with silky hairs, pileipellis often extending and exceeding lamellae. Context thin, white. Lamellae free, white at first, then gradually changing to pale pink, close. Stipe 5–15 cm long, 1–2.5 cm thick, subcylindrical, tapering upwards, white, glabrous, solid, slightly curved. Volva large and thick, saccate, white to pale yellow or dirty brown, with cracks or tomentose scales. Basidiospores 6.5–10 × 4.5–6.5 μm, broadly elliptical to oval, smooth, pale pink.

In summer to autumn, solitary or gregarious on rotten wood of broad-leaved trees. Edible; cultivable. Distributed in Northeast, Northwest, Northern, and Central China.

1555 *Volvariella gloiocephala* (DC.) Boekhout & Enderle

Pileus 6–13 cm broad, campanulate at first, then gradually plane, prominent at centre, glabrous, viscid, pinkish grey brown to lotus-root pink, brownish grey at centre, with long stripes or ribs at margin. Context white to dirty white. Lamellae free, greyish white at first, then gradually changing to light cinnamon, subclose. Stipe 7–17 cm long, 1–1.5 cm thick, cylindrical, gradually enlarged towards to base, white or lighter than pileus, solid to soft insides. Volva white, cupulate. Basidiospores 10–14.5 × 7–8 µm, broadly elliptical to elliptical, smooth, pale pink.

In summer to autumn, solitary or gregarious on ground in grassland or broad-leaved forests. Edible, but also reported as toxic, please be careful when eating it. Distributed in Northeast, Central, Southern, and Northwest China.

1556 *Volvariella hypopithys* (Fr.) Shaffer

Pileus 3–4 cm broad, campanulate to hemispherical at first, convex when expanded, often prominent at centre, white to dirty white, with pink tinge of basidiospore colour with age, with long fibrils, inconspicuously to obviously striate. Context white to dirty white. Lamellae free, subclose, white, pinkish fleshcolour to pink, unequal. Stipe 1–5 cm long, 2–3 mm thick, cylindrical, white, solid to soft insides, enlarged at base. Volva white, subsaccate to cupulate. Basidiospores 5.5–9 × 4–6.5 µm, elliptical, smooth, pale pink.

In summer to autumn, solitary to scattered on grassland or ground in forests. With both edible records and poisonous records, it is not recommended.to eat it. Distributed widely in China.

1557 *Volvariella nivea* T.H. Li & Xiang L. Chen

Pileus 7–9 cm broad, subconical at first, then plane to a convex, pure white, not viscid, even at margin, thin, estriate. Context 4–5 mm thick at stipe, thin, white, unchanging when injured, with mild odour. Lamellae 5–7 mm broad, free, close, with 8–9 lamellae per cm at pileus margin, white when young, changing to pink when mature. Stipe 10–11.5 cm long, 0.7–0.8 cm thick, cylindrical, slightly silky striate, white. Volva fleshy, saccate, white. Basidiospores 6–7 × 4.5–5.5 μm, oval to broadly elliptical, smooth, pale pink.

On ground in bamboo or broad-leaved forests. Distributed in Southern China.

1558 *Volvariella pusilla* (Pers.) Singer

Pileus 0.5–5 cm broad, oval at first, gradually changing to campanulate to hemispherical, then plane and prominent at centre, white to dirty white, dry, with silky hairs, even to striate at margin at first. Context thin, white. Lamellae free, white to pink. Stipe 1–5 cm long, 2–5 mm thick, cylindrical, white, glabrous, solid. Volva membranous, often lacerated at margin. Basidiospores 5–6.5 × 4–5 μm, oval to subglobose, smooth, pale pink.

In summer to autumn, solitary or gregarious on grassland, or on ground in parks or forests. Edible. Distributed in Northern, Northwest, and Central China.

1559 *Volvariella subtaylor* Hongo

Pileus 3–5 cm broad, flat to plane, slightly prominent at centre, dirty white, covered with brown to grey subradially appressed fibrillose scales. Context white. Lamellae free, pale pink. Stipe 5–8 cm long, 3–6 mm thick, white to beige, tomentulose to subglabrous. Volva 1–1.5 cm long, 0.8–1.2 cm broad, cupulate, brown to grey brown, thin, with whitish tomentum. Basidiospores 6–7.5 × 4–5 μm, broadly elliptical to oval, smooth, pale pink.

In summer to autumn, on ground in mixed forests. Distributed widely in China.

1560 *Volvariella volvacea* (Bull.) Singer

Pileus 10 cm broad, 5 mm thick, greyish white to dark grey when fresh, usually darker at centre, gradually lighter at margin, radially striate, grey brown when dried; margin sharp, incurved when dried. Context 2 mm thick, pale yellow when dried, soft corky. Lamellae close, unequal, free, cream at first, then pink, yellow brown when dried. Stipe 7–9 cm long, 0.5–2 mm thick, cylindrical, white, glabrous, fibrous, solid, pale yellow when dried, brittle. Volva 5 cm broad, cupulate, cream to grey black. Basidiospores 7.5–8.5 × 5–6 μm, elliptical to broadly elliptical, smooth, pale pink, inamyloid.

In summer to autumn, on strawstack and grassland rich in organic matters. Edible and medicinal. Cultivable. Distributed in Central and Southern China.

1561 *Xeromphalina campanella* (Batsch) Kühner & Maire

Pileus 1–3 cm broad, hemispherical at first, depressed to umbilicate at centre, then nearly infundibuliform when margin expanded, moist, glabrous, orange yellow, obviously striate at margin. Context thin, membranous, yellow. Lamellae adnate to decurrent, pale yellow to pale orange, close to subdistant, unequal, widish, interveined. Stipe 1–4 cm long, 2–5 mm thick, cylindrical, often tapering downwards, pale yellow brown in upper portion, dark red brown in lower portion, soft to hollow insides. Basidiospores 6–7.5 × 2–3.5 µm, elliptical, smooth, hyaline, amyloid.

In summer to autumn, gregarious on decayed stumps in forests. Edible. Distributed in Northeast, Northern, Northwest, Central, and Southern China.

1562 *Xeromphalina cauticinalis* (Fr.) Kühner & Maire

Pileus 2.5 cm broad, convex to plano-convex or plane, shallowly depressed at centre, glabrous, widely radially striate or wrinkled at margin, orange brown to yellow brown, darker at centre. Context thin, yellow, brittle. Lamellae decurrent, distant, interveined, pale yellow. Stipe 2–8 cm long, 0.1–0.2 cm thick, cylindrical, enlarged at base, pale yellow on upper portion, light red brown to dark brown in lower portion, hairy, or subglabrous on upper portion, with rust hairs at base. Basidiospores 4–7 × 2.5–3.5 µm, elliptical, smooth, hyaline, amyloid.

In summer to autumn, solitary, scattered or gregarious on debris and rotten wood in coniferous forests. Distributed in Northwest China.

1563 *Xeromphalina tenuipes* (Schwein.) A.H. Sm.

Pileus 3–6 cm broad, oblate hemispherical to plane, sometimes slightly depressed at centre, apricot yellow, yellow to honey yellow, darker at centre, radially sulcate, with brown tomentum but gradually glabrescent. Context thin, yellow. Lamellae sinuate, pale yellow to yellow brown, distant. Stipe 4–7 cm long, 0.3–1 cm thick, cylindrical, covered with yellow or yellow brown (on upper half portion) to dark brown (on lower half portion) tomentum. Basidiospores 7–9.5 × 4.5–5 µm, elliptical to long ellipsoid, smooth, hyaline, amyloid.

In summer, on rotten wood in tropical and subtropical forests. Distributed in Southern China.

1564 *Xerula sinopudens* R.H. Petersen & Nagas.

Pileus 1–4.5 cm broad, oblate hemispherical to convex, depressed at centre, pale grey, pale brown to yellow brown, densely covered with grey brown to brown bristles. Context thin, white to greyish white. Lamellae sinuate or adnate, white to beige, distant. Stipe 3–10 cm long, 3–5 mm thick, cylindrical, covered with brown bristles. Basidiospores 10.5–13.5 × 9.5–12.5 µm, subglobose to broadly elliptical, smooth, hyaline. Pleurocystidia thin-walled, not encrusted with crystals.

In summer, on ground in tropical and subtropical forests. Edible. Distributed in Southern and Central China.

1565 *Xerula strigosa* Zhu L. Yang et al.

Pileus 2–5 cm broad, oblate hemispherical to convex, yellow brown, dark brown to grey brown, densely covered with yellow brown bristles. Context 4 mm thick, whitish. Lamellae sinuate to adnate, white to beige, subdistant. Stipe 5–10 cm long, 3–6 mm thick, covered with yellow brown bristles. Basidiospores 11–15 × 9–11.5 μm, broadly elliptical to elliptical. Pleurocystidia thick-walled, encrusted with crystals at apex.

In summer, on ground in subtropical to temperate forests. Edible. Distributed in Central China.

Afterword

Atlas of Chinese Macrofungal Resources is about to be published soon! How excited we all are—just like a mother expecting a baby to be born, looking forward to the publication of this book, for it is the fruit of the author's many years of hard work.

The names of the co-editors of this book are arranged in strokes by the last name. Yucheng Dai provided photos of polypores, which are nearly 1/3 of the entire book; Tolgor Bau and Taihui Li provided pictures of agarics and ascomycetes; Zhuliang Yang provided pictures of amanitas and boletes, including a large number of new species. Meanwhile, Li Fan, Zengzhi Li, Haixia Ma, etc. (the rest not listed) provided fine items in their respective research fields. Taihui Li put in a lot of hard work in the process of collating the text of the book, especially in the final stage.

China is the country with the richest biodiversity in the northern hemisphere, and the macrofungi in it are even more abundant and colourful. But what group do macrofungi with various features exactly belong to? This is a huge mystery that has drawn us to explore without ceasing. After decades of research, we have picked up a little. Compared to the types of macrofungi that actually exist in nature, the 1817 species introduced in this atlas are not many. However, works that introduce more than 1800 species of macrofungi with both fine pictures and texts are not common in the world, and it is unprecedented in China! Why is such a work so rare? There are many reasons for this, but the authors believe that the most important reason is: how difficult it is!

First of all, the accurate identification of fungi is not easy. Although science and technology are advancing by leaps and bounds and the world seems to be omnipotent; human understanding of fungi is still limited, and there are many species that have never been discovered or have been misidentified in the past! Therefore, identifying each species correctly is a big challenge to mycologists, and sometimes it is quite difficult to identify even the most common "general species"! If time and space went back to before 2000 AD, when some people would say that the black wood-ear that is often consumed in China was the European wood-ear *Auricularia auricula-judea* (Bull.) Quel., yuanmo mushroom (winter mushroom) consumed with large volume in Northeast China the European oyster mushroom *Hohenbuehelia serotina* (Pers.) Singer, the "white boletes" sold in large quantities in market in Yunnan during the rainy season *Boletus edulis* Bull., the most commonly cultivated Lingzhi the European *Ganoderma lucidum* (Curtis) P. Karst., the most common deadly poisonous white poisonous mushroom ("Destroying Angel of Guangzhou") in Southern China the European *Amanita verna* (Bull.) Lam., etc., almost nobody would doubt the identification results of these "general species". However, systematic research by the authors of this book shows sufficient scientific evidence to prove that these identifications are indeed mistaken! They are actually native to China (or Asia), and their correct scientific names should be: *Auricularia. heimuer* F. Wu et al., *Panellus edulis* Y.C. Dai et al., *Boletus. meiweiniuganjun* Dentinger, *Ganoderma lingzhi* Sheng H. Wu et al. and *Amanita. exitialis* Zhu L. Yang & T.H. Li. Re-recognizing a common species or verifying a new species is certainly not as easy as telling a lion from a tiger. It usually takes months or even years or longer from the discovery and verification of a new species to the publication of a paper. For example, our understanding of *Ganoderma leucocontextum* T.H. Li et al. has gone through a long process of more than 20 years—one of the authors collected the specimens of this species for the first time in Sichuan Province in the 1980s and "naturally" identified it as "*Ganoderma lucidum*". After in-depth observation and comparison to European specimens in the early 1990s, he began to doubt the original identification results. However, it was not until recent years that the ideal specimens were collected again in Tibet and Sichuan with collaborators and after sufficient morphological and molecular systematic evidence was obtained, was it finally proven to be a new species in 2014. There are more than 230 species in this book that were first discovered and published by the research team led by the five main authors of the book. Had there not been long-term accumulation, it would have been difficult to have such a large number of first-hand research materials. Nobody dares to jump to conclusions easily while identifying a common "general species", not to mention the greater difficulties while verifying a new species. The workload and the difficulty of authenticating are evident.

© Central China Farmer's Publishing House 2025

Y. Li et al., *Atlas of Chinese Macrofungal Resources*, https://doi.org/10.1007/978-981-96-0502-6

Anyone who is engaged in the investigation and research of biological resources in the field will most likely encounter various difficulties and obstacles—the discomfort of being exposed to the sun and rain; the danger of steep mountains and deep valleys; the harassment by midges, rats and ants, the poison of snakes, scorpions, bees and ticks, etc. are beyond words. But it is worth mentioning that it is particularly difficult to take photos of macrofungi in the wild. Many macrofungi prefer to grow in dark, humid and dense forests and the collection period is often in the rainy season, which makes it inconvenient for photographing. In addition, swarms of mosquitoes, flies, locusts or poisonous insects are often annoying, leaving no peace for taking ideal photos; what's more, many macrofungi are only suitable for photographing during a short period of time, and photos of rare species are even harder to come by. The photos used in this book are entirely attributed to the authors' years of accumulation and the selfless dedication by colleagues and friends.

From the beginning of writing to the publishing of this book, it took the authors more than 3 years, during which, in addition to completing their own teaching and research tasks, the authors had to make time to complete the writing, integrating, editing, proofreading and revising of manuscripts. The tasks are far more arduous than originally imagined. In order to pursue scientific accuracy, artistic layout, and the readability of text, the authors and the relevant staff of the publishing house worked together avidly for countless days and nights. This book was not completed until being revised again and again. While this book is soon to be published, we would like to take this opportunity to express our gratitude to the colleagues who supported us and the leaders, editors, printing typesetters and proofreaders of Central China Farmer's Publishing House!

Mankind's understanding of macrofungal resources continues, and the road of scientific exploration is far from ending. It is not our hope to take this book as a summary of the authors' decades of work, but rather a new stage for exploring the diversity of China' s fungal resources in the future, and a new beginning for a higher goal. We believe that in the near future, there will be more new discoveries of macro fungal resources in the country.

Due to the rapid development of fungal taxonomy in recent years, the frequency of changes in the scientific names and taxonomic status of fungi is also accelerating. In addition, due to the limitations of the author's ability and other factors, unintentional roughness, errors and omissions in the book are inevitable, and with the improvement of understanding of science as time goes, the shortcomings will become more obvious. Therefore, we sincerely welcome comments and corrections from our readers. We sincerely hope that more people with lofty ideals will join this field and contribute to the improvement of the level of mycological research in the country. We are also willing to work together with all people to open the maze of "China's fungal resource varieties" and make unremitting explorations and forever enterprising.

The authors
July 2015

Further Reading

Banerjee P, Sundberg WJ. The genus *Pluteus* section *Pluteus* (Pluteaceae, Agaricales) in the midwestern United States [J]. Mycotaxon. 1995;53:189–246.

Bau T, Bao HY, Li Y. A revised checklist of poisonous mushrooms in China [J]. Mycosystema. 2014;33(3):517–48.

Bau T, Liu Y. Three new records of the omphalinoid fungi in China [J]. Mycosystema. 2010;29(5):767–70.

Bau T, Yevgeniya MB, Zhuang JY, et al. Agarics and other macrobasidiomycetes from Ussuri River Valley [J]. Mycosystema. 2007;26(3):349–68.

Bau T, Zhang H. New records of *Galerina* collected from Changbai Mts. in China [J]. Mycosystema. 2012;31(1):55–61.

Bau T. Colourful world of mushrooms [M]. Shanghai: Shanghai Science Popularization Press; 2012.

Bau T. Flora fungorum Sinicorum Vol. 49 Strophariaceae [M]. Beijing: Science Press; 2014.

Bi ZS, Li TH, Zhang WM, et al. A preliminary agaric flora of Hainan Province [M]. Guangzhou: Guangdong Higher Education press; 1997.

Bi ZS, Li TH, Zhen GY. Two new species of the genus *Gymnopilus* Karsten [J]. Mycosystema. 1986;5(2):93–8.

Bi ZS, Li TH. A preliminary note on *Russula* species from Guangdong, with a new species and a new variety [J]. Guihaia. 1986;6(3):193–9.

Bi ZS, Zhen GY, Li TH. Marcofungus flora of Guangdong [M]. Guangzhou: Guangdong Science & Technology Press; 1994.

Bi ZS, Zhen GY, Li TH. Marcofungus flora of the mountainous district of north Guangdong [M]. Guangzhou: Guangdong Science & Technology Press; 1990.

Comprehensive Scientific Expedition Team to Qinghai-Tibet Plateau of Chinese Academy of Sciences. Fungi of Tibet [M]. Beijing: Science Press; 1983.

Dai XC, Li TH, Li WF, et al. Macrofungus flora of Garze Prefecture, Sichuan Province [M]. Chengdu: Sichuan Science and Technology Press; 1994.

Dai YC, Bau T. Illustrations of edible and medicinal fungi in Northeastern China [M]. Beijing: Science Press; 2007.

Dai YC, Li Y. Notes on the nomenclature of six important medicinal fungi in China [J]. Mycosystema. 2011;30(4):515–8.

Dai YC, Yang ZL. A revised checklist of medicinal fungi in China [J]. Mycosystema. 2008;27(6):801–24.

Dai YC, Zhou LW, Yang ZL, et al. A revised checklist of edible fungi in China [J]. Mycosystema. 2010;29(1):1–21.

Deng CY, Li TH, Li T, et al. New species and new Chinese records of *Marasmius* sect. *Sicci* [J]. Cryptogam Mycol. 2012;33(4):439–51.

Deng CY, Li TH, Song B. A new species and a new record of *Marasmius* from China [J]. Mycotaxon. 2011;116:341–7.

Deng CY, Li TH. *Gloeocantharellus persicinus*, a new species from China [J]. Mycotaxon. 2008;106:449–53.

Deng CY, Li TH. *Marasmius galbinus*, a new species from China [J]. Mycotaxon. 2011;115:495–500.

Deng CY, Li TH. *Marasmius nigrodiscus*, a newly recorded species to China [J]. Mycosystema. 2008;27(5):768–70.

Deng WQ, Li TH, Chen NZ, et al. A critical note on the scientific name of the cultivated edible fungus, Zhudugu [J]. Acta Edulis Fungi. 2006;13(3):71–4.

Deng WQ, Li TH, Li P, et al. A new species of *Amanita* section *Lepidella* from South China [J]. Mycol Prog. 2014;13:211–7.

Deng WQ, Li TH, Shen YH. A new species of *Clitopilus* from southwestern China [J]. Mycotaxon. 2012;122:443–7.

Deng WQ, Li TH, Shen YH. A study of the types and additional materials of *Clitocybe pseudophyllophila* and *Clitocybe subcandicans* [J]. Mycotaxon. 2008;103:377–80.

Deng WQ, Li TH, Wang CQ, et al. A new crepidotoid *Entoloma* species from Hainan Island (China) [J]. Mycoscience. 2015;56(3):340–4.

Deng WQ, Shen YH, Li TH. A small cyathiform new species of *Clitopilus* from Guangdong, China [J]. Mycosystema. 2013;532(5):781–4.

Fan YG, Bau T, Takahito KY. Newly recorded species of *Inocybe* collected from Liaoning and Inner Mongolia [J]. Mycosystema. 2013;32(2):302–8.

Fan YG, Bau T. *Inocybe hainanensis*, new lilac-stiped species from tropical China [J]. Mycosystema. 2014;33(5):954–60.

Fan YG, Bau T. Two striking *Inocybe* species from Yunnan Province, China [J]. Mycotaxon. 2013;123(13):169–81.

Fujian Sanming Regional Fungal Experiment Station. Fujian fungus illustration (I) [G]; 1973.

Fujian Sanming Regional Fungal Experiment Station. Fujian fungus illustration (II) [G]; 1978.

Fungi Group, Institute of Microbiology, Chinese Academy of Sciences. Poisonous mushroom [M]. Beijing: Science Press; 1975.

Ge ZW, Yang ZL, Vellinga EC. The genus *Macrolepiota* (Basidiomycota) in China [J]. Fungal Divers. 2010;45:81–98.

Ge ZW, Yang ZL, Zhang P, et al. *Flammulina* species from China inferred by morphological and molecular data [J]. Fungal Divers. 2008;32:59–68.

Guo QX, Fan YG, Bau T. Newly recorded species of *Inocybe* from Jilin Province in China [J]. Mycosystema. 2014;33(1):162–6.

Hao YJ, Qin J, Yang ZL. *Cibaomyces*, a new genus of Physalacriaceae from East Asia [J]. Phytotaxa. 2014;162(4):198–210.

Hawkworth DL, Kirk PM, Sutton BC, et al. Ainsworth & Bisby's dictionary of the fungi [M]. 8th ed. Wallingford: CAB International; 1995.

He XL, Horak E, Li TH, et al. Two new cuboid-spored species of *Entoloma* s. l. (Agaricales, Entolomataceae) from southern China [J]. Cryptogam Mycol. 2015;36(2):237–50.

He XL, Li TH, Jiang ZD, et al. *Entoloma mastoideum* and *E. praegracile*—two new species from China [J]. Mycotaxon. 2011;116:413–9.

He XL, Li TH, Jiang ZD, et al. Four new species of *Entoloma* s.l. (Agaricales) from southern China [J]. Mycol Prog. 2012;11:915–25.

He XL, Li TH, Jiang ZD, et al. Type studies on four *Entoloma* species from South China [J]. Mycotaxon. 2012;121:435–45.

He XL, Li TH, Jiang ZD. Three species of white *Entoloma* new to China [J]. Mycosystema. 2010;29(6):920–3.

© Central China Farmer's Publishing House 2025
Y. Li et al., *Atlas of Chinese Macrofungal Resources*, https://doi.org/10.1007/978-981-96-0502-6

He XL, Li TH, Peng WH, et al. A taxonomic revision of *Entoloma subclitocyboides* and *E. subinfundibuliforme* from Hainan Island, South China [J]. Mycoscience. 2013;55(2):103–7.

He XL, Li TH, Xi PG, et al. Phylogeny of *Entoloma* s.l. subgenus *Pouzarella*, with descriptions of five new species from China [J]. Fungal Divers. 2013;58:227–43.

He XL, Peng WH, Gan BC. Morphological and molecular evidence for a new species in *Entoloma* subgenus *Claudopus* from Sichuan Province, southwest China [J]. Mycoscience. 2015; https://doi.org/10.1016/j.myc.2014.10.001.

He XL, Ye XJ, Li TH, et al. New and noteworthy species of white *Entoloma* (Agaricales, Entolomataceae) in China [J]. Phytotaxa. 2015;205(2):99–110.

Horak E. Agaricales of New Zealand 1: Pluteaceae-Entolomataceae. Fungi of New Zealand, vol. 5 [J]. Fungal Divers Res Ser. 2008;19:1–305.

Huang NL. Colored illustrations of macrofungi (mushrooms) of China [M]. Beijing: China Agriculture Press; 1998.

Huang NL. Edible fungi cyclopedia [M]. Beijing: China Agriculture Press; 1997.

Justo A, Malysheva E, Bulyonkova T, et al. Molecular phylogeny and phylogeography of Holarctic species of *Pluteus* section *Pluteus* (Agaricales: Pluteaceae), with description of twelve new species [J]. Phytotaxa. 2014;180(1):1–85.

Justo A, Minnis AM, Ghignone S, et al. Species recognition in *Pluteus* and *Volvopluteus* (Pluteaceae, Agaricales): morphology, geography and phylogeny [J]. Mycol Prog. 2011;10(4):453–79.

Justo A, Vizzini A, Minnis AM, et al. Phylogeny of the Pluteaceae (Agaricales, Basidiomycota): taxonomy and character evolution [J]. Fungal Biol. 2011;115(1):1–20.

Kirk PM, Cannon PF, David JC, et al. Ainsworth & Bisby's dictionary of the fungi [M]. 9th ed. Wallingford: CAB International; 2001.

Kirk PM, Cannon PF, Minter DW, et al. Dictionary of the fungi [M]. 10th ed. Wallingford: CAB International; 2008.

Li JZ, Hu XW, Peng YB. Macrofungus flora of Hunan Province [M]. Changsha: Hunan Normal University Press; 1993.

Li RG. Economic macrofungi of northeast China [M]. Changchun: Northeast Normal University Press; 1998.

Li RG. Edible, medicinal and poisonous mushrooms of northeast China [M]. Changchun: Northeast Normal University Press; 1992.

Li RG. Flora fungorum of Jilin, vol. 1, Basidiomycotina [M]. Changchun: Northeast Normal University Press; 1991.

Li RG. Toxic and harmful fungi of Jilin Province [M]. Changchun: Jilin People's Press; 1980.

Li TH, Chen XL, Shen YH, et al. A white species of *Volvariella* (Basidiomycota, Agaricales) from southern China [J]. Mycotaxon. 2009;109:255–61.

Li TH, Deng WQ, Deng CY, et al. A new species of *Lentinellus* from China [J]. J Fungal Res. 2012;10(3):130–2.

Li Y, Bau T. Flora fungorum sinicorum, vol. 45, Pleurotoid—lentinoid fungi [M]. Beijing: Science Press; 2014.

Li Y, Bau T. Mushrooms of Changbai mountains, China [M]. Beijing: Science Press; 2003.

Liu B. Chinese medicinal fungi [M]. Taiyuan: Shanxi People's Press; 1978.

Liu B. Edible macrofungi in Shanxi [M]. Taiyuan: Shanxi United University Press; 1991.

Liu PG, Yuan MS, Wang XH, et al. Notes on the resources of matsutake group and their reasonable utilization as well as effective conservation in China [J]. J Nat Resour. 1999;14(3):245–52.

Liu Y, Bau T, Li TH. Three new records of the genus *Hohenbuehelia* in China [J]. Mycosystema. 2010;29(3):454–8.

Liu Y, Bau T. A new species of *Hohenbuehelia* from China [J]. Mycotaxon. 2009;108(2):445–8.

Liu Y, Bau T. Two new species of *Lentinellus* (Russulales, Basidiomycota) in China [J]. Mycosystema. 2011;30(5):680–5.

Mao XL, Jiang CP, Tsewang U. Economic macrofungi of Tibet [M]. Beijing: Science Press; 1993.

Mao XL. Economic fungi of China [M]. Beijing: Science Press; 1998.

Mao XL. Macrofungi in China [M]. Zhengzhou: Henan Science and Technology Press; 2000.

Mao XL. Macromycetes of China [M]. Beijing: Science Press; 2009.

Mao XL. Poisonous mushroom identification [M]. Beijing: Popular Science Press; 1987.

Minnis AM, Sundberg WJ. *Pluteus* section *Celluloderma* in the U. S. A. [J]. N Am Fungi. 2010;5(1):1–107.

Neda H. Correct name for "nameko" [J]. Mycoscience. 2008;49(1):88–91.

Orton PD. British fungus flora, agarics and boleti 4: Pluteaceae: *Pluteus* and *Volvariella* [M]. Edinburgh: Royal Botanic Garden; 1986.

Pegler DN. A preliminary agaric flora of East Africa [J]. Kew Bull Addit Ser. 1977;6:1–615.

Pegler DN. Agaric flora of Lesser Antilles [J]. Kew Bull Addit Ser. 1983;9:1–668.

Pegler DN. Agaric flora of Sri Lanka [J]. Kew Bull Addit Ser. 1986;12:1–519.

Pradeep CK, Justo A, Vrinda KB, et al. Two new species of *Pluteus* (Pluteaceae, Agaricales) from India and additional observations on *Pluteus chrysaegis* [J]. Mycol Prog. 2012;11(4):869–78.

Qin J, Feng B, Yang ZL, et al. The taxonomic foundation, species circumscription and continental endemisms of *Singerocybe*: evidence from morphological and molecular data [J]. Mycologia. 2014;106(5):1015–26.

Qin J, Hao YJ, Yang ZL, et al. *Paraxerula ellipsospora*, a new Asian species of Physalacriaceae [J]. Mycol Prog. 2014;13(3):639–47.

Ren W. Forest diseases in Yunnan, China [M]. Kunming: Yunnan Science and Technology Press; 1993.

Shao LP, Xiang CT. Forest mushrooms of China [M]. Harbin: Northeast Forestry University Press; 1997.

Singer R. Monographs of South American basidiomycetes, especially those of the east slope of the Andes and Brazil: 1. The genus *Pluteus* in South America [J]. Lloydia. 1958;21:195–299.

Tai FL. Sylloge fungorum sinicorum [M]. Beijing: Science Press; 1979.

Teng SC. Fungi of China [M]. Beijing: Science Press; 1963.

The Edible Fungi Research Institute of Shanghai Academy of Agricultural Sciences. Edible fungal flora of China [M]. Beijing: China Forestry Publishing House; 1991.

Tian EJ, Bau T. New records of *Pholiota* subgen. *Pholiota* in China [J]. Mycosystema. 2013;32(5):907–12.

Tian EJ, Bau T. *Pholiota virescens,* a new species from China [J]. Mycotaxon. 2012;121:153–7.

Tian EJ, Bau T. *Stropharia jilinensis*, a new species (Strophariaceae, Agaricales) from China [J]. Nova Hedwigia. 2014;99(1/2):271–6.

Tian EJ, Bau T. Two newly recorded species of *Pleuroflammula* from China [J]. J Northeast For Univ. 2011;39(9):128–9.

Tong CL, Bao HY, Bau T. Chemical constituents and antibacterial activity of petroleum ether extract from fruit bodies of *Tricholoma mongolicum* [J]. Mycosystema. 2010;29(4):619–24.

Vellinga E, Yang ZL. *Volvolepiota* and *Macrolepiota-Macrolepiota velosa*, a new species from China [J]. Mycotaxon. 2003;85:183–6.

Vizzin A, Perrone L, Gelardi M, et al. A new collection of *Chlorophyllum hortense* (Agaricaceae, Agaricales) from southeastern China: molecular confirmation and morphological notes [J]. Riv Micol Romana. 2014;91(1):3–19.

Wang CQ, Li TH, Huang H, et al. *Marasmius albopurpureus*, a new species of section *Globulares* from Baili Island, China [J]. Mycol Prog. 2015; https://doi.org/10.1007/s11557-015-1053-z.

Wang CQ, Li TH, Song B. *Hygrocybe griseobrunnea*, a new brown species from China [J]. Mycotaxon. 2013;125:243–9.

Wang CQ, Li TH, Zhang M, et al. A new species of *Hygrocybe* subsect. *Squamulosae* from South China [J]. Mycoscience. 2015;56:345–9.

Wang JR, Bau T. A new species and a new record of the genus *Entoloma* from China [J]. Mycotaxon. 2013;124:165–71.
Wang L, Yang ZL, Liu JH. Two new species of *Laccaria* (Basidiomycetes) from China [J]. Nova Hedwigia. 2004;79(3/4):511–7.
Wang XH, Liu PG, Yu FQ. Color atlas of Wild commercial mushroom in Yunnan [M]. Kunming: Yunnan Science and Technology Press; 2004.
Wang XH, Yang ZL, Li YC, et al. *Russula griseocarnosa* sp. nov. (Russulaceae, Russulales), a commercially important edible mushroom in tropical China: mycorrhiza, phylogenetic position, and taxonomy [J]. Nova Hedwigia. 2009;88(1/2):269–82.
Wei TZ, Yao YJ, Li TH. First record of *Termitomyces bulborhizus* in China [J]. Mycotaxon. 2003;88:433–8.
Wu XL, Dai YC, Li TH, et al. Fungi of tropical China [M]. Beijing: Science Press; 2011.
Xie ZX, Wang Y, Wang B. Illustrations of agarics in Changbai Mountains, China [M]. Changchun: Jilin Science and Technology Press; 1986.
Xu JT. Chinese medicinal mycology [M]. Beijing: Beijing Medical University, China Union Medical University Press; 1997.
Yang SS, Bau T, Li TH. New Chinese records of *Pluteus* collected from Jilin Province, China [J]. Mycosystema. 2011;30(5):794–8.
Yang SS, Bau T. Three new records of *Crepidotus* from northern China [J]. Nova Hedwigia. 2014;98(3/4):507–13.
Yang XF, Li FQ, Han SL, et al. Study on the resources of medicinal macrofungi in the Dabie Mountain of Henan [J]. Plant Sci J. 2005;23(4):393–7.
Yang YZ. Prevention and treatment manual of poisonous fungi [M]. Beijing: People's Medical Publishing House; 1983.
Yang ZL, Feng B, Hao YJ. *Pseudoarmillariella bacillaris*, a new species with bacilliform basidiospores in Asia [J]. Mycosystema. 2013;32(Suppl):127–32.
Yang ZL, Feng B. The genus *Omphalotus* (Omphalotaceae) in China [J]. Mycosystema. 2013;32(3):545–56.
Yang ZL, Li TH, Wu XL. Revising of *Amanita* collections from Hainan, Southern China [J]. Fungal Divers. 2001;6:149–65.
Yang ZL, Li TH. Notes on three white *Amanitae* of section *Phalloideae* (Amanitaceae) from China [J]. Mycotaxon. 2001;78:439–48.
Yang ZL, Li YC, Tang LP, et al. *Trogia venenata* (Agaricales), a novel poisonous species which has caused hundreds of deaths in southwestern China [J]. Mycol Prog. 2012;11(4):937–45.
Yang ZL, Weiss M, Oberwinkler F. New species of *Amanita* from the eastern Himalaya and adjacent regions [J]. Mycologia. 2004;96(3):636–46.
Yang ZL, Zhang LF, Mueller GM, et al. A new systematic arrangement of the genus *Oudemansiella* s. str. (Physalacriaceae, Agaricales) [J]. Mycosystema. 2009;28(1):1–13.
Yang ZL. Die *Amanita*-Arten von Südwestchina [J]. Bibl Mycol. 1997;170:1–240.
Yang ZL. Flora fungorum sinicorum Vol. 27 Amanitaceae [M]. Beijing: Science Press; 2005.
Ying JZ, Mao XL, Ma QM. Atlas of Chinese medicinal fungi [M]. Beijing: Science Press; 1987.
Ying JZ, Zang M. Economic macrofungi in southwest China [M]. Beijing: Science Press; 1994.
Ying JZ, Zhao JD, Mao XL, et al. Edible mushroom of China [M]. Beijing: Science Press; 1982.
Yuan MS, Sun PQ. Sichuan mushroom [M]. Chengdu: Sichuan Science and Technology Press; 1995.
Yunnan Health and Epidemic Prevention Station. Common edible and poisonous fungi in Yunnan [M]. Kunming: Yunnan People's Publishing House; 1961.
Zhang H, Bau T. Two species of *Tubaria* (Agaricales) new to China [J]. Mycosystema. 2010;29(4):588–91.
Zhang P, Chen ZH, Xiao B, et al. Lethal amanitas of East Asia characterized by morphological and molecular data [J]. Fungal Divers. 2010;42(1):119–33.
Zhang ST, Mao XL. Hong Kong mushrooms [M]. Hong Kong: The Chinese University Press; 1995.
Zhang WM, Li TH, Bi ZS, et al. Taxonomic studies on the genus *Entoloma* from Hainan Province of China (I) [J]. Mycosystema. 1994;13(3):188–98.
Zhang WM, Li TH, Bi ZS, et al. Taxonomic studies on the genus *Entoloma* from Hainan Province of China (II) [J]. Mycosystema. 1994;13(4):260–3.
Zhao Q, Zhang Y, Yuan LC. Study on the resources and utilization of medical fungi in Laojun Mountain [J]. J Microbiol. 2006;26(4):85–8.
Zhao ZY, Mao XL. Atlas of macrofungi in Xinjiang, China [M]. Urumqi: Xinjiang Bayi Agricultural College; 1986.

GPSR Compliance

The European Union's (EU) General Product Safety Regulation (GPSR) is a set of rules that requires consumer products to be safe and our obligations to ensure this.

If you have any concerns about our products, you can contact us on ProductSafety@springernature.com

In case Publisher is established outside the EU, the EU authorized representative is:

Springer Nature Customer Service Center GmbH
Europaplatz 3
69115 Heidelberg, Germany

Batch number: 10756367

Printed by Printforce, the Netherlands